POLLINATION

The Enduring Relationship between Plant and Pollinator

POLLINATION

The Enduring Relationship between Plant and Pollinator

TIMOTHY WALKER

Princeton University Press

Princeton and Oxford

Published in the United States and Canada in 2020 by Princeton University Press,
41 William Street, Princeton, New Jersey 08540
press.princeton.edu

Copyright © 2020 Quarto Publishing plc

Conceived, designed and produced by The Bright Press,
an imprint of The Quarto Group
The Old Brewery, 6 Blundell Street,
London N7 9BH, United Kingdom
T (0) 20 7700 6700 F (0)20 7700 8066
www.QuartoKnows.com

All Rights Reserved

ISBN: 978-0-691-20375-1

Library of Congress Control Number: 2020936816

This book has been composed in Archer and Effra
Printed on acid-free paper

Written by Timothy Walker
Publisher: James Evans
Editorial Director: Isheeta Mustafi
Managing Editor: Jacqui Sayers
Senior Editor: Caroline Elliker
Editorial Assistant: Chloe Porter

Art Director: Katherine Radcliffe

Commissioned by Jacqui Sayers

Project Manager: David Price-Goodfellow, D & N Publishing
Designer: Studio Noel
Illustrations: Robert Brandt

Printed in Singapore

10 9 8 7 6 5 4 3 2 1

Contents

Introduction 6

Chapter 1
What is Pollination? 8

Chapter 2
Wind and Water 44

Chapter 3
Animals 66

Chapter 4
Attraction 96

Chapter 5
Rewards 126

Chapter 6
The Biological Significance of Pollination 154

Chapter 7
Pollination, Pollinators and People 182

Chapter 8
Conservation of Pollinators 198

Bibliography 216
Index 217
Picture credits 224

PREVIOUS PAGE
Honeybee (*Apis mellifera*) visiting apple (*Malus* sp.) blossom.
RIGHT
Hoverfly on an aster (*Symphyotrichum* sp.) bloom.

Introduction

Pollination is an essential component of the world's ecosystems. The process is required for the successful reproduction of nine out of every ten plants on Earth, and without it the world would revert to the state it was in 350 million years ago. This world would certainly not contain humans.

Pollination is both simple and complex. In simple terms, it is the transport of a pollen grain from the male parts of a flower to the female part of a flower. However, that transport is facilitated by a wide variety of animals, the wind and occasionally water. And the conversation between the pollen grain and the female part of a flower is one of the most extraordinary in nature and has no equivalent in human biology.

The relationships between pollinators and flowers are among the most intimate in nature. For example, some orchid flowers produce a scent that mimics the alarm call of a bee under attack, in order to attract a hornet that subsequently pollinates the orchid. In other species the flower scent may be retch-inducing to humans but attracts pollinating carrion beetles and flies.

The rewards or payments handed out by flowers include scents that the pollinator can subsequently use to attract a mate for itself. Other flowers are used by the pollinators as mating arenas and brood chambers. Whenever pollination is studied, sex, lies and deception are rarely far away.

Pollination has been studied for centuries by botanists and zoologists alike, perhaps most famously by the naturalist Charles Darwin (1809–1882), and research in the area is currently expanding at a rate never previously witnessed. However, despite hundreds of years of study and countless scientific papers on this subject, pollination is still an area of fierce debate. The cherished belief that by looking at a flower you can deduce its pollinator is clearly untenable.

At the moment we are realising how little we know about pollination, we are also realising that pollinators are declining around the world, for myriad reasons. Concern about the status of pollinators has reached governments and the United Nations, and their conservation is now the subject of international programmes, with ambitious targets set for 2030. This book explains why hitting these targets is critical, not only in ecological terms, but also for ethical and economic reasons.

Tansy (*Tanacetum vulgare*) flowers attract honeybees.

Chapter 1

What is Pollination?

Pollination is a uniquely plant event, involving the transfer of a pollen grain from the male structure of a plant to a female structure. There is no equivalent in the animal kingdom, and yet for 87.5 per cent of angiosperms the process involves an animal. Like many other close and intimate relationships, pollination can be good, fair and equally beneficial for both partners. However, selfishness can triumph, and deceit, lies and false promises can evolve and succeed. To understand pollination fully, we must explore not only where it fits into the life histories of plants that produce seeds, but also what comes before and after the event.

Common blue butterfly (*Polyommatus icarus*).

The global relevance of pollination

One thing all plants that produce pollen have in common is that they produce seeds. A seed is basically an embryo protected by a coat and sustained by a food supply. The significance of the evolutionary emergence of seeds should not and cannot be underestimated, as these structures are remarkable survival capsules. The record length of time for a seed to remain alive and subsequently grow into a flowering plant is 30,000 years, for a seed from an immature *Silene* fruit found in Siberia. Seeds are even a means for plants to survive fires, which have been a common occurrence since at least the Devonian period 385 million years ago. Seeds are also a means of dispersing the embryos of the next generation. In addition to being a highly significant evolutionary innovation for plants, seeds have become a vital component of the lives of many animals, including *Homo sapiens*. Because of the great importance of seeds, it is vital that we understand and, if necessary, protect the processes that lead to their production. Pollination is one of these processes.

FOSSIL FIND
A coloured scanning electron micrograph (magnification: ×1,600) of a fossilised pollen grain from the extinct conifer genus *Corollina* or *Classopollis*. A relative abundance of these pollen grains marks the end of the Triassic, around 195 million years ago.

Scientists don't know exactly when plants first produced pollen grains. This is a tricky date to pin down for at least three reasons: the fossil record is incomplete and unevenly distributed in time and space; a fossil pollen grain can closely resemble a fossil spore from a fern; and pollen grains are very small. Nor do we know whether pollen grains evolved before seeds, again because of the incomplete nature of the fossil record. The group of plants that are misleadingly referred to as seed ferns was already established in the early Carboniferous period 360 million years ago. These plants had leaves like ferns, but they produced seeds and so were not ferns. They are, however, all extinct and so are slightly enigmatic. The best-known fossil of a pollen-bearing structure belongs to another extinct species, *Wielandiella angustifolia*. Its pollen looks very similar to that found in Triassic fossils from 200 million years ago. It appears that seeds pre-date pollen. Whatever the sequence of events, pollen grains and seeds together have contributed to the domination of the terrestrial environment by seed-bearing plants, which in turn support the majority of land-based organisms living today.

FIRST POLLEN PRODUCERS
Some of the oldest fossils of a pollen-producing reproductive structure belong to plants in the family Williamsoniaceae. Often called an inflorescence or even flower, this was actually a female or bisexual cone surrounded by bracts.

Sexual reproduction in plants

This story starts with the evolution of sexual reproduction in plants more than 500 million years ago, at which time all plants lived in water. We now know that sexual reproduction has actually evolved at least three times, once in each of the major kingdoms: plants, fungi and animals. Research has shown that lineages (populations or entire species) that reproduce only clonally, or asexually, do not last as long as those that reproduce sexually. Asexual reproduction may be good in the short term if the plant is exceptionally well adapted to its habitats or if there is no other organism with which it can breed, but sexual reproduction is better in the longer term.

There is no single reason why sexual reproduction is better than asexual reproduction. In fact, it can even be seen as disadvantageous because it breaks up a winning combination of genes – it is unlikely that the children of an Olympic athlete will be as fast or as strong as their triumphant parent. However, sexual reproduction does allow for the possibility of an even more successful combination of genes and therefore traits. In the animal kingdom there is evidence that sexual reproduction is very useful for introducing disease resistance into a population, and it also seems to have increased the rate at which new species can emerge. This is particularly important if we assume that high species diversity is preferable to low species diversity. In addition, there is good evidence that sexual reproduction is the reason why, in a changing environment, populations of organisms that reproduce in this way survive longer than those that reproduce asexually.

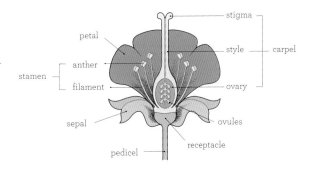

BASIC FLOWER ANATOMY

Although flowers vary hugely, most retain four standard components: sepals, petals, stamens and carpels. The green sepals protect the developing flower in bud, while the colourful petals attract pollinators. Stamens produce pollen, which is transferred to the carpel's stigma, from where genetic material travels into the ovary.

ASEXUAL REPRODUCTION

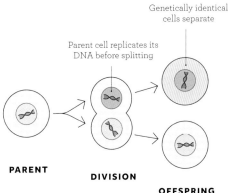

SEXUAL REPRODUCTION

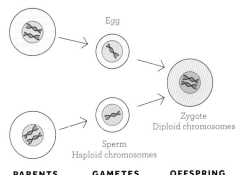

Other problems of sexual reproduction are that having unisexual males and females means that each individual can mate with only half of the population, and that only females produce offspring. Because most plants are bisexual, this is less of a problem for them than it is for animals. That said, being bisexual brings other problems, including self-fertilisation (selfing) and inbreeding depression, or reduced biological fitness (see page 36).

The reason for describing the advantages of sexual reproduction in detail here is that it shows just how important it is for the long-term survival of a population of a species. In turn, this means that all the individual components of sexual reproduction – including pollination – are very important.

Before pollen and seeds

The fact that life evolved in water is not surprising. Water is a benign habitat whose temperature fluctuates slowly, it provides some protection from harmful ultraviolet light, and gas exchange between the living cells of organisms and the water works well. The water in which early plants grew contained nutrients excreted by the animals with which they shared their environment. Furthermore, the water provided support and so the plants did not need to synthesise energy-costly woody tissues. Finally, the water was a medium through which plant sperm could swim in search of an appropriate egg.

However, about 470 million years ago a plant managed to survive on dry land away from a permanent film of water for at least part of the year. This move was fraught with problems, all of which had to be solved over the next many millions of years. It should be noted here that during evolution organisms sometimes 'reassign' a feature that evolved for one reason but turns out to be useful for another. These types of characteristics are known as exaptations, a word coined in 1982 by palaeontologists Stephen Gould and Elisabeth Vrba to replace the term pre-adaptation, which erroneously implies that evolution can have foresight. After leaving the water, plants had to solve the immediate problem of desiccation. However, the issue in terrestrial plants of preventing water from leaving through evaporation is similar to that faced by aquatic plants in preventing excess water entering cells through osmosis. As result, a thick cuticle will do the job in both a marine or terrestrial environment. In addition, gas exchange, nutrient uptake and protection from ultraviolet light on land all needed to be sorted out, followed later by the issues associated with a taller growth form.

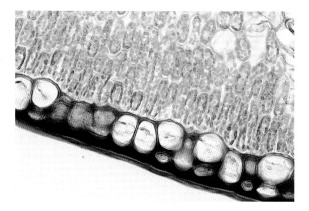

ALL-IMPORTANT CUTICLE
The cuticle on the leaves of land plants is essential to prevent harmful desiccation.

All of these problems required modifications and adaptations that permitted plants to survive out of the water and in the air on land. But survival was not enough – these intrepid colonisers also needed to reproduce, which meant that they had to find a way for their sperm to continue to meet eggs. They solved this by continuing to use water through which the sperm could swim.

Initially, this meant that sexual reproduction was restricted to those times of the year when the plants were covered in a film of water. This clearly was not a game-ending problem – in fact, mosses, liverworts, hornworts, lycopods and ferns today all still reproduce at those times of year when there is water on their surface. For these plants, the requirement for a film of water to enable sexual reproduction has not been a problem for 470 million years. It is useful at this point to look in detail at the life history of ferns, as it helps to explain how pollination evolved and also how the role of water has changed in the process yet remained critically important.

TREES OF LIFE
Evolutionary trees, or phylogenies, are a widely used way to demonstrate the relationships between groups of organisms, in this case land plants. However, they are constantly being refined and there are still arguments over how these plants are related to one another, in particular the bryophytes.

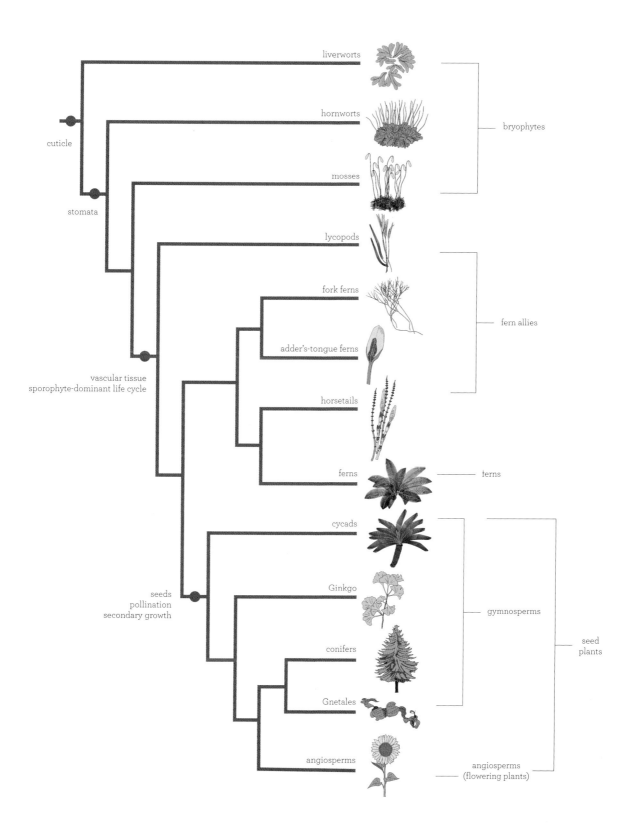

SEXUAL REPRODUCTION IN PLANTS 13

Focus on ferns

On the underside of the fronds of most ferns you can often see brown rust-like structures called sporangia (singular sporangium). These produce dust-like spores, which are released on dry days or when you put a frond lower side down on a piece of paper in a dry place. The patterns of the sporangia on fern fronds are sometimes very useful in distinguishing the species, as well as being very beautiful. The spores themselves are single-celled entities wrapped up in an amazing material called sporopollenin (see box on page 22).

The spores are released into the air and blown away from the parent plant, hopefully landing near spores from unrelated plants of the same species (generally speaking, inbreeding is undesirable in plants, as it is in animals – see page 36). When the spore lands in a place that is damp and warm enough, it will absorb water and start to grow. If there is no soil and no shade where it has germinated, it will soon die. But if it has landed on a damp patch of soil in the shade, it will grow into a small flat green plant the size of a baby's fingernail.

This little plant, known as a prothallus, contains the organs that produce the eggs and the sperm – the gametes. Because the prothallus produces gametes it is called a gametophyte (the suffix '-phyte' is derived from the Greek word for plant). The organs that produce the male gametes, or sperm, are called antheridia (singular antheridium), and the organs that produce the female gametes, or eggs, are called archegonia (singular archegonium). The antheridia each produce many sperm, which are released into the film of water covering the prothallus. Each archegonium produces just one egg, which stays within the structure waiting for the sperm to come to it. The sperm each have several flagella to propel them through the water, but how do they 'know' the direction in which they should swim? There is good evidence that the archegonia send out a chemical attractant signal, and this may well involve calcium. The flagella not only propel the sperm but are also used to steer them, although the details of how they achieve this are still largely unknown.

FERN SPORANGIA
On the underside of fern fronds you can often see the spore-producing sporangia, known as sori. The patterns are often beautiful and also very useful in species identification.

The sperm typically swim a few centimetres, although distances of up to 1 m (3 ft) have been recorded. Even so, this is not very far if their goal is to find an egg on an unrelated prothallus. Unlike mosses (see box), ferns do not appear to have evolved any mechanisms to transport their sperm further afield.

ABOVE: LIVERWORTS

The common or umbrella liverwort (*Marchantia polymorpha*; above left) is frequently seen growing in plant containers. Its male and female gametes (sperm and eggs) are produced underneath the umbrella-like structures. The sperm swim from beneath the disc-shaped structures (above right) to the eggs, under the star-shaped structures (right).

AN EARLY ASSOCIATION

Mosses seem to benefit from the services of microarthropods such as springtails (shown here) and mites for transporting their sperm. Mosses pre-date ferns and this interaction with animals could pre-date the insect-mediated pollination of seed-bearing plants by tens of millions of years. It might also explain why there are nearly twice as many species of moss as there are of ferns: the greater stirring of the gene pool in the mosses could lead to higher rates of speciation, as could the differential movement of microarthropods. The behaviour of these animals could isolate the gene pools of populations, causing them to drift apart genetically resulting in the emergence of new species.

When the sperm does find an egg from a plant of the same species, it fuses with it. Because each egg and each sperm is haploid, with just one set of chromosomes and hence one copy of each gene, the cell that results from their fusion is diploid, with two sets of chromosomes and two copies of each gene. There are several advantages to diploidy, one of which is that if one copy of a gene is damaged or corrupted, then this broken gene will be masked and the good gene will effectively cover for it. (You could equally say that this is a disadvantage in so far as damaged genes can persist, hidden in the population.)

The new diploid cell is called a zygote. In the human life history, the zygote is also the first diploid cell formed by the fusion of two gametes, each of which is haploid. In the life history of the fern, the embryo that develops from the zygote is dependent on the prothallus for its nutrition, in the same way that a mammal embryo depends on its mother. But here the analogy ends, because the young fern will take from its maternal prothallus until the latter has no more to give and dies. By the time this happens, the young diploid fern is self-supporting and free-living, with its own 'root' system and photosynthetic fronds. After a few years the plant will become mature, and when this occurs the new fronds will produce sporangia, in which the spores develop (for this reason, the fern plant is called the sporophyte). These spores are released to disperse to pastures new and the fern life cycle is complete.

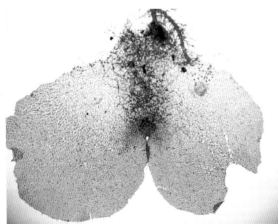

FERN PROTHALLUS
This tiny structure produces sperm and eggs.

UNFURLING FROND
Bracken (*Pteridium aquilinum*) fronds emerge looking like a clenched fist, but they subsequently unfurl to give the more familiar mature structure.

A critical difference

The typical fern life history described above seems a bit odd to us because it involves two different free-living plants, the tiny prothallus (the gametophyte) and the familiar and much bigger fern plant (the sporophyte). But having two life stages that appear completely different is not actually that strange – it happens in butterflies and moths, for example, with their hungry caterpillars and more beautiful mature flying stages. However, one crucial fact was missing from the summary of a fern life history, which is central to understanding the true nature of pollination and the pollen that is being transported.

In the life history of every sexually reproducing organism there has to be a moment when haploid cells are produced. If this did not happen, the number of copies of each gene would double every time a zygote is formed. In humans, when sperm are formed in the testes of a male or eggs are produced in the ovary of a female, the number of chromosomes in each cell is halved during the process of meiosis. It may be tempting to think that things are the same in plants, and that meiosis takes place in their antheridia and archegonia, but temptation can sometimes lead you astray. In fact, meiosis in terrestrial plants happens in the sporangia, and the spores, covered in their protective coat of sporopollenin, are actually haploid.

This can be a bombshell for anyone who has an entrenched view that gametes are always produced immediately after meiosis. While this may be true for animals, it is not the case in plants. Here, the spores that are produced immediately after meiosis grow into a multicellular haploid stage, which then produces the sperm and eggs. In ferns, this multicellular haploid stage is the tiny, free-living prothallus, which is the gamete-producing gametophyte. In some fern species the prothallus develops both sperm-producing male antheridia and egg-producing female archegonia, whereas in others the prothallus is either male or female. In gymnosperms (including conifers) and angiosperms (flowering plants), the multicellular stage that produces the gametes is always a different structure and the sperm is produced in the haploid pollen grain.

PLANT LIFE HISTORY

Diploid Multicellular Plant
Meiosis + differentiation
↓
Haploid Unicellular Spores
Mitosis + differentiation
↓
Haploid Multicellular Plant
Mitosis + differentiation
↓
Haploid Unicellular Gametes
Fusion of 2 gametes
↓
Diploid Unicellular Zygote
Mitosis + differentiation
↓
Diploid Multicellular Plant
↓

The production of spores by meiosis and gametes by mitosis is not found in the animal life history. In animals, there are no spores, and gametes are produced by meiosis thus:

ANIMAL LIFE HISTORY

Diploid Multicellular Animal
Meiosis + differentiation
↓
Haploid Unicellular Gametes
Fusion of 2 gametes
↓
Diploid Unicellular Zygote
Mitosis + differentiation
↓
Diploid Multicellular Animal

What is pollen?

The pollen produced by seed-bearing plants such as gymnosperms and angiosperms is an immature male gametophyte. This means that the pollen derives from the production of a haploid spore in a sporangium. In gymnosperms, the sporangia that produce the pollen grains are in small, soft cones near the tips of new shoots. In angiosperms, the sporangia are in the anthers at the end of filaments that make up the stamens found in the flowers, often near the petals.

Generally speaking, gymnosperm pollen is made up of four cells, although this can vary between species. In contrast, angiosperm pollen is made up of two cells, or one cell with two nuclei. This then begs the questions: how are the sperm produced? And where is the antheridium? The answer is that during evolution the antheridium was found to be surplus to requirement and pollen grains could produce sperm more effectively without them. In angiosperm pollen grains one of the cells is known as the generative cell, and it is this cell that divides to produce two sperm. Once this has happened, the pollen grain is a mature male gametophyte. But we are getting ahead of ourselves here, because the sperm are not produced until the pollen has arrived close to a female gametophyte of the same species. This raises a further question: where is the female gametophyte?

In gymnosperms and angiosperms the egg-producing female gametophyte does not leave the plant that produces the spore from which it developed. In gymnosperms this is generally a woody cone and in flowering plants it is inside the ovary in the centre of the flower. Because the female gametophyte does not leave the plant, the male gametophyte has to come to her. It is this journey made by pollen that is at the heart of pollination, and it ends when the pollen grain arrives either on a female gymnosperm cone or on the stigmatic surface of an angiosperm flower.

LEFT: CLOUDS OF POLLEN

When plants employ wind to transport their pollen, they have to produce up to millions of grains to guarantee that enough of them find a female structure on a plant of the same species.

OPPOSITE: THE BEAUTY OF POLLEN

Photographs taken with a scanning electron microscope reveal the beauty of the patterns on the outside of pollen grains.

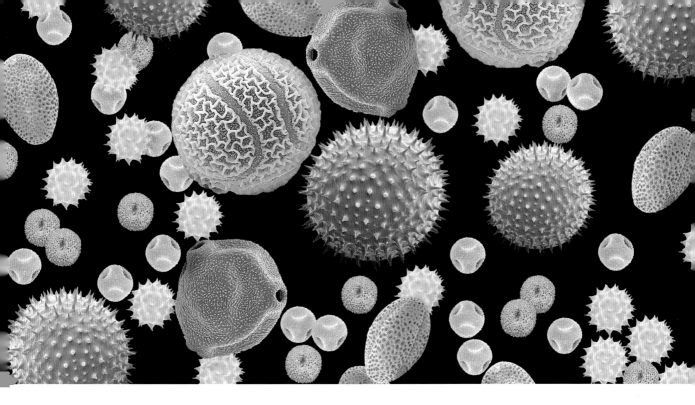

What does a pollen grain look like?

Pollen grains are small, but they vary considerably in size, as well as in a number of other ways. They range in diameter from 4 μm to 350 μm, averaging 35 μm, and their volume varies by five orders of magnitude (1 μm, or micrometre, is a thousandth of a millimetre). It is one thing for the dimensions to vary across the 360,000 species of seed plants, but there is also environment-influenced variation within individual species.

The patterning on the outside of pollen grains is extraordinarily variable – and also very beautiful. In addition, the colour and scent of pollen differ from species to species. These last two features may be connected with attracting animals, but some of the scents are produced by volatile oils, and these may have functions unrelated to the manipulation of a pollinator. For example, some volatiles have antibiotic properties and are thus protective, while others taste or smell unpleasant and so dissuade herbivores from eating the pollen (pollen that is eaten will never realise its potential as a supplier of sperm). In some cases, volatile chemicals from one grain have been shown to affect another pollen grain's ability to fulfil its function. This is referred to as allelopathy and can be a serious problem. And finally, the evaporation of volatiles can be a means of heat loss, thus preventing pollen grains from overheating.

Despite the significant variation among the pollen grains produced by seed plants, there are few clear correlations between the form of the pollen and the pollinator, the environment or the taxonomic group to which the plant belongs. The pollen grains of members of some plant families do have shared characteristics – species in the gourd family (Cucurbitaceae), for example, have larger grains, while those in the borage family (Boraginaceae) and have smaller grains, as is the case with most wind-pollinated plants. However, pollen generally tells us very little about the life of the plant from which it came. One very important thing that it can tell us is the identity of the plant, as many pollen grains can be identified even to species level. This, coupled with the fact that pollen grains make very resilient fossils and have survived as such unscathed for thousands and millions of years, makes them indispensable to palaeobotanists and biogeographers, among many others (see Chapter 6).

Pollen grain structure

The basic structure of a pollen grain is a thick, protective coat, a supply of nutrients for the contents, and the cells that are going to produce the sperm and ultimately feed them. The grain must protect and support its contents during its journey, during which it faces several serious hazards. First, it is at risk of desiccation. Because pollen grains are tiny, their surface area to volume ratio is very high and loss of water would be rapid were the coat not supremely waterproof. Second, exposure to sunshine can result in the pollen getting very hot. Evaporation of volatile substances has already been mentioned as a means of losing heat, but the coat may be reflective or simply act as an insulation layer. Another problem that results from exposure to the sun is damage to genetic material from ultraviolet light. A further threat faced by pollen grains in transit comes from fungi and bacteria that may infect and kill them.

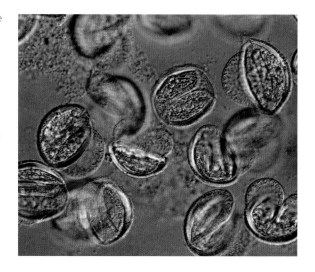

BUOYANCY TANKS
The pollen grains of pine trees have balloon-like air sacs to help keep them airborne long enough to reach another plant of the same species.

POLLEN GRAIN

There is a lot going on inside a pollen grain. The vegetative cell is responsible for delivering the sperm, and the generative cell is responsible for producing the sperm.

The pollen coat

Aside from providing protection, the pollen coat has other roles to play once the grain reaches its destination in the female cone of a gymnosperm or on the female part of a flower. The first of these is adhesion and the second is recognition of the plant upon which the pollen has landed or been deposited. The structure of the pollen coat has several distinct layers: the outer pollenkitt, the exine, the intine and the normal cell membranes around the cells in the pollen grain.

Pollenkitt

The outermost layer is called the pollenkitt. It contains lipids that have reflective properties to help prevent the pollen grain from overheating, and it can also contain adhesives that are useful for sticking the pollen to the vector and then to the female part of the flower. The pollenkitt layer is sometimes involved in recognition and rejection systems to reduce the risk of self-pollination. Unfortunately, it is also the part of the pollen grain that can cause the allergic reactions referred to as hay fever.

Exine

The second layer is the exine, which is mostly made up of sporopollenin (see box on page 22). The thickness of the layer and its internal structure vary greatly, sometimes along taxonomic lines. The sporopollenin gives the pollen grain its external appearance in terms of the surface sculpturing. Counterintuitively, the pattern of the sculpturing appears to have nothing to do with the pollinator, nor with the surface upon which it will land, and the attractive hypothesis of a molecular lock-and-key docking mechanism has failed to receive support. The thicker the exine, the more robust the pollen grain. In some cases this does seem to be linked to the finesse of the pollinator. For example, pollen moved by perching birds seems to be more robust than that moved by the more delicate hummingbirds.

The primary function of the exine could therefore be seen as protection, with a common secondary role being the recognition that the pollen grain is at the correct destination. The molecular mechanism of this recognition involves not only the exine itself but also the cell membrane.

However, the pollen grain needs to be able to take up water, germinate and produce a tube (the pollen tube) that will grow through the female part of the plant to the egg-producing female gametophyte. In order to grow this tube, there has to be at least one pore or aperture in the pollen wall. In the case of flowering plants, there are either one or three apertures.

INSIDE THE GRAIN
A single pollen grain contains important genetic material wrapped up inside four protective layers.

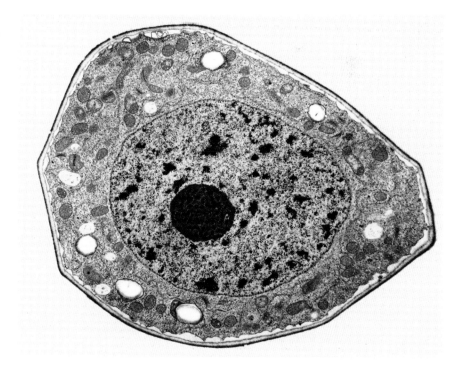

Sporopollenin

This remarkable substance gives pollen grains their resilience to physical, chemical and biological attack.

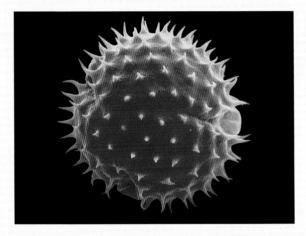

The word sporopollenin implies that the substance has something to do with spores (such as those released from the fronds of ferns) and pollen, and indeed it does: it is found in the protective coats of both. It is also found in the coating of some algal spores. It is clear, therefore, that sporopollenin is an exaptation (see page 12). It (or a chemical with similar properties) was a prerequisite for the successful colonisation of land. So is pollen a spore? The answer is, not really. Pollen is a male gametophyte in spore's clothing. Pollen is a heavily armoured sperm factory and delivery system, although it is a sperm factory that produces just two sperm.

Despite the importance of sporopollenin, its chemical structure remained unknown until 2019, when a group of chemists and biologists from the Massachusetts Institute of Technology and the Whitehead Institute for Biomedical Research, both in Cambridge, Massachusetts, reported that pine sporopollenin is mostly made up of 'aliphatic-polyketide-derived polyvinyl alcohol units and 7-*O*-*p*-coumaroylated C16 aliphatic units, crosslinked through a distinctive dioxane moiety featuring an acetal. Naringenin was also identified as a minor component of pine sporopollenin.' This description conveys something of the complexity of the material – the diagram of the chemical structure is reminiscent of the external skeletons of some modern city office blocks! The scientists speculate that with this knowledge the way may now be open to designing and synthesising highly resistant materials for industrial purposes.

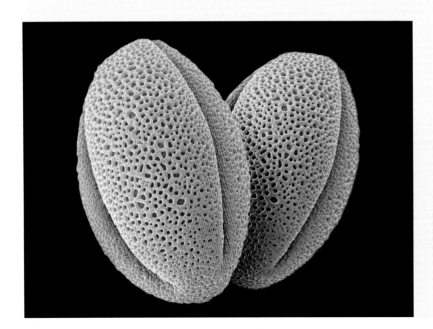

SPOROPOLLENIN SURFACES

Although they are both composed of sporopollenin, the pollen grains of marigolds (above) and peonies (left) have different surfaces.

Intine
The third layer of the pollen coat is basically a normal plant cell wall made up of a mixture of hemicelluloses, cellulose and pectin. It is referred to as the intine.

Cell membrane
The fourth and final layer of the coat is the normal cell membrane surrounding the cells in the pollen grain.

The pollen interior

The cells contained within a pollen grain include many of the normal organelles and other bits and pieces. However, they contain an above-average amount of starch and sugars, which provide sustaining energy for the pollen grain's journey and extra energy for the development of the two sperm when the time comes.

Pollen faces one particular problem as a direct result of the starch and sugars it contains to nourish the male gametophyte, a problem shared by seeds, which contain food to nourish the embryo. That is, whenever a plant has evolved an energy store for its own use, some wretched animal eventually comes along and eats it. As already stated, digested pollen grains are of no use for delivering gametes. When faced with this dilemma, whether it concerns seeds that are being eaten or pollen grains, one solution is for plants to produce seeds or pollen in sufficient quantities to satiate the animals that feed on them, with some left over to fulfil their function and continue the species.

DEVASTATING PEST
The UK harvest of rapeseed oil was seriously reduced in 2019 due to the consumption of pollen by the common pollen beetle (*Meligethes aeneus*).

Pollen longevity

Despite the features described above, pollen is not immortal; nothing on Earth is. It is actually very difficult to measure how long pollen can remain viable. In the right conditions, such as controlled temperature and humidity, it can survive for weeks or even months, which is very useful for plant breeders. However, some pollen is very short-lived. Included in this group are the grasses, and thus the cereals upon which we depend for our staple food. These pollen grains remain viable for only a few hours or even minutes. However, this is not a problem in a farmer's field, where the plants are packed in so tightly that pollen journey times are very short.

There is a correlation between the availability of a pollinator and longevity of pollen. When pollinators are scarce, such as in the winter, pollen seems to be longer lived. Plants that are more tolerant of self-fertilisation tend to have shorter-lived pollen, perhaps because the pollen has to take a very short journey. Conversely, where plants are widely distributed at a low density, such as epiphytic orchids growing on the branches of trees in tropical forest, pollen is usually able to survive for longer. It is now known that, on average, wind-dispersed pollen can live for just 25.5 hours while insect-dispersed pollen may survive for 8.5 days.

LEFT: APPARENT OVERPRODUCTION

Grasses are notorious for producing large quantities of pollen.

OPPOSITE: STRANGE ORCHIDS

By contrast, orchids produce tiny seeds that resemble dust.

In summary, many factors influence pollen longevity. These include the morphology of the flower and its position on the plant; the environmental conditions throughout the whole process, from the production of the spore through to its arrival on another cone or flower; and the morphology of the pollen grain.

If, by now, you are beginning to feel that there are few (if any) rules governing the variation in pollen grains, then you would be right. Even within the relatively small group of gymnosperms (about a thousand species) there is an extraordinary number of variations. It appears that pollen grains are not well designed for a narrow range of conditions and for a small guild of vectors, and that being a jack of all trades is a better strategy than becoming a specialist. This might be a clear indication that pollination is a peril-filled period in the life of plants, and that they can best ensure their success by producing pollen that can cope with a wide variety of these threats and challenges.

The stages of pollination

At this point in the story it is important to clarify exactly when pollination begins and when it ends. Different biologists have different opinions on what constitutes pollination. This is not surprising, because essentially successful life is a chain of individual lives, each made up of chains of events. In this book, pollination is considered to be the event of a plant's life that starts at the moment a viable pollen grain leaves the structure wherein it has been produced. And pollination ends at the moment the pollen grain has germinated successfully in the right place and is actively growing towards the egg, which will then be fertilised by the sperm produced in the pollen grain.

The release of the pollen is possible only after it has been constructed. Likewise, the arrival of pollen on the stigma and its growth down the style has a point only if the ovary and its ovules have developed at the base of that style. The production of the pollen in the anthers and of the ovules in the ovary are not described in this book, because the start and finish lines of pollination have to be drawn somewhere.

The event known as pollination can be broken down into several stages:

1. Release of the pollen by the male cone or anther.
2. Presentation of the pollen by the flower.
3. Removal of the pollen by the vector (see Chapters 2 and 3).
4. Transport of the pollen (see Chapters 2 and 3).
5. Navigation of the vector (see Chapters 4 and 5).
6. Deposition of the pollen on a female cone or stigmatic surface.
7. Successful germination of the pollen grain and growth towards the egg.

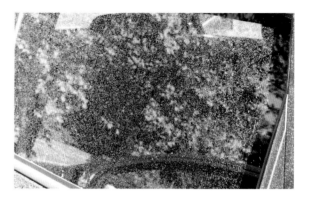

POLLEN SCREEN
A car parked under a cedar tree on a still night in the spring is liable to be heavily dusted with pollen grains come morning.

1. Release of the pollen

If you park your car under a cedar (*Cedrus* sp.) tree on a mild, dry night in spring, you will return in the morning to a car covered in yellow dust resembling custard powder. Or, in similar weather, try tapping the branches of a pine tree (*Pinus* sp.) or yew (*Taxus* sp.) when the 1 cm-long (0.4 in) soft male cones are visible – again, a cloud of dust will appear. In both cases, the dust-like substance is pollen. The important point to note here is that the pollen is released when the weather is dry, as this desiccates the cone, freeing the grains.

If you are tempted to smell a lily (*Lilium* sp.) flower and you push your nose into the centre of the petals, you may be rewarded by a dab of bright orange pollen on your nose. The significance of the colour will be considered in Chapter 4, but the point here is that lily pollen is released in one go when the anther splits all the way down one side. In flowering plants, the pollen is produced in, and released from, the anthers on the end of filaments. There are many variations in the way the pollen is released, however, and not every flower behaves in the same way as lily flowers.

As the pollen grains develop within the anther, they are surrounded by a nutrient-rich fluid. The soup is then either reabsorbed by the anther as the grains mature or is allowed to evaporate. This facilitates the opening of the anther, either by a longitudinal split as in lilies, or by

the appearance of circular pore at the end of the anther, as seen in members of the heath family (Ericaceae). Simultaneously, the pollen grains dry out (by as much as 80 per cent) and so are in a dormant state when they start their journey.

The timing of the opening can vary. Sometimes it can be controlled by the plant to coincide with the activity of pollinators – for example, bee-pollinated flowers tend to open their anthers in the morning and moth-pollinated flowers tend to open theirs towards the end of the day. Some flowers open all their anthers simultaneously, while others open them sequentially, thereby prolonging the period during which the flower is in the male stage. Geranium (*Geranium* spp.) flowers have two rings of five stamens, with one of the rings maturing and dehiscing (opening) before the other. In other species the plant does not control the process and the anthers simply open when the environment has dried them out. Some flowers that open their anthers following desiccation even reseal them if the humidity rises.

When whole plants rather than individual flowers are observed, further variation is found in the pattern of pollen release. In a single foxglove plant (*Digitalis* sp.), for example, the flowers will be in a wide range of maturity at any one time, from closed to anthers fully open, right through to maturing fruits.

POLLEN ON THE LOOSE!

Pollen can be released in different ways, depending on the species concerned. Rhododendrons (top) have holes or pores at the tip of their anthers, out of which the pollen is shaken, several grains at a time. Lilies, on the other hand (bottom), have anthers that split vertically and bend backwards, revealing all of their pollen in one go.

SIMULTANEOUS POLLEN RELEASE

Work carried out in the 1990s by Graham Stone and colleagues showed that in some African acacias (*Acacia* spp.) all flowers on all plants of the same species at a particular location open simultaneously. It has been suggested that this is a means of optimising pollination success, promoting outcrossing and reducing the loss of pollen through herbivory by satiating the pollinators. This might be risky if there are no pollinators in the vicinity when the trees are in flower, but researchers observed that the different species release their pollen at different times of the day. Even though these trees are using the same pollinators (mainly bees), their pollen is moved mostly to other members of the same species because the pollinators move rapidly from tree to tree. The different acacia species happily coexist by avoiding competing for the pollinators at the same time of day, and hence there is no daily spike in pollen release.

2. Presentation of the pollen

The majority of pollen is passively displayed in the flower, from where it is removed either actively or passively by the vector (see Chapters 2 and 3, which examine the different vectors involved). In several species, however, the pollen is given a helping hand to start it on its way. One of these is creeping dogwood (*Cornus canadensis*), which opens its flowers in 0.5 milliseconds, allegedly making them the fastest-moving plant structure in the world. At the exact moment the pollinator lands, pollen is flung out of the flower by a mechanism reminiscent of a medieval trebuchet. One feels a certain sympathy for the poor fly that is attracted to the inflorescence by its white bracts, only to be pebble-dashed with supersonic pollen. Interestingly, this is just one of a growing number of plant species so far discovered that use both animals and/or wind as a vector.

FORCIBLE EVICTION

A tachinid fly visiting the flowers of creeping dogwood (*Cornus canadensis*). The flower between the fly's centre and rear legs has sprung open, violently firing pollen onto the body of the unsuspecting insect.

The most violent pollen ejection is possibly that of the Australian trigger plants (*Stylidium* spp.). The stamens of these little plants are fused to the style in a column reminiscent of (but different from) that seen in orchids. In the set position this column is held behind the petals, but when the insect lands on the flower it swings out of hiding like a hammer, hitting the insect with a shower of pollen. The anthers then wither as the hammer is reset. When the next insect arrives, the receptive stigma is hammered onto the visitor, which hopefully has brought pollen from another plant.

A less violent shaking of pollen is achieved in the beautiful flowers of kalmias (*Kalmia* spp.), where stamens snap out of little pockets in the petals as the pollinator pushes into the flower. Sage (*Salvia* spp.) flowers have just two fertile stamens, counterbalanced on a short filament. These swing on a fulcrum onto the pollinator as it pushes into the flower looking for nectar. A more gentle approach is taken by mahonia (*Mahonia* spp.) flowers, where the filaments bend inwards, gently stroking the anther against the pollinator.

A final example of proactive presentation of pollen, known as the plunger mechanism, is found in the daisy family (Asteraceae). In the flowers of these species, five stamens are arranged in a ring around the central carpel (the female part of the flower, consisting of the stigma atop the style, which connects the stigma to the ovary and its ovules). The sausage-shaped anthers are fused to form a cylinder and split open on their inner surface, releasing their pollen into the centre of the cylinder.

The pollen is then pushed out of the cylinder by the style as it grows up through the cylinder. Cleverly, the two-lobed style is shut tight. Only once the style has cleared the anthers does the stigma open to produce the characteristic Y-shaped structure. The pollen is pushed out and away from the receptive surface of the stigma, thereby reducing the likelihood of self-pollination.

PLUNGER MECHANISM

In the flowers of plants in the daisy family (Asteraceae), the pollen is often pushed off the anthers as the style grows up through the centre of the flower.

A FERN-LIKE FEATURE

The pollen in the male flowers of castor oil plants (*Ricinus communis*) is blown out by the release of mechanical tension in the anthers, which builds up as the anthers dry out. This piece of bioengineering is analogous to that seen in ferns, where the spores are flung away from the frond by the release of tension following desiccation of the sporangia. Although the male flowers actively expel their pollen for only one day, the female flowers are receptive for ten days, meaning that they are likely to receive pollen from another plant and not just from their own male flowers.

3. Removal of the pollen by the vector

Different vectors remove pollen in different ways. The transfer may be the result of deliberate action by the vector, or just a happy accident facilitated by the design of the flower. This is covered in more detail in Chapters 2 and 3.

4. Transport of the pollen

The ways in which pollen is transported varies from the totally stochastic, uncontrolled blowing-in-the-wind method, to the very careful carriage exemplified by yucca moths in the family Prodoxidae. Again, these different means of transport are looked at in Chapters 2 and 3. Whatever the vector, pollen grains have evolved to be well prepared for this journey, from their resilient coat to their in-flight nutrition (see page 21).

5. Navigation of the vector

The pollen grain is a vehicle without a driver. In most cases it is a passenger on an animal, but as such it cannot even contribute to the navigation of its journey. Instead, this is the responsibility of both the flower within which the pollen is produced and the flower to which it wants to be taken, plus of course a sentient pilot to pick up the clues in the landscape. The pollen-producing flower must flag down the bus and put the pollen on board safely and securely, and then the receptor flower (preferably not on the same plant but in the same species) must flag down the same bus and enable the pollen to alight. If the plant is wind pollinated, then the operation is much simpler: the pollen just jumps and hopes that it lands somewhere suitable. This is dealt with in more detail in Chapters 4 and 5.

6. Deposition of the pollen

Alighting from the vector at the correct place is another perilous moment for the pollen grain, and sometimes the pollinator goes out of its way to make life difficult for the plant. For example, a honeybee (*Apis mellifera*) collecting pollen for the nest will store it in pollen baskets (corbiculae), thenceforth making it unavailable to the flowers. The only pollen that can be transferred to a stigma is that which falls on the parts of the bee it cannot groom, such as the middle of its abdomen. At other times the pollinator is very cooperative. The classic example of this is the yucca moth, which deliberately places pollen on the stigmatic surface. For gymnosperms, the vector is most often wind, and in most species a pollination drop plays part of the role of the stigma – the pollen just has to land in the fluid (see page 33 for more on this).

Actually landing on the stigma is a much more complicated stage in the journey than it initially appears. If there were just one species of plant and just one pollinator, then all journeys from one plant to another could potentially be successful. However, there are many species of seed plants (more than 360,000) and many pollinators (potentially more than 1.2 million). The pollen may be put on the wrong plant, or it may not even get off the pollinator because it is buried under a layer of pollen from another species. In this scenario the pollen could simply miss its destination. The problem can be eliminated if the plant and animal evolve a close working relationship that excludes any other species, such as that seen in yuccas (*Yucca* spp.) and their moths (see page 151), but this can create other problems, such as the absence – even temporary – of either party.

OPPOSITE: INSECT VECTOR
While bees may carry loads of pollen in 'baskets' on their legs, the grains usually gather loosely all over the insect; these are thought to be the pollen that does much of the pollination in plants.

MULTI-TASKING BEES

It is well known that some vectors can carry pollen from more than one orchid species at the same time, but on different parts of their body that reflect the flower structure of the species concerned. For example, in South Africa the long-legged oil-collecting bee *Rediviva longimanus* has been seen (shown here) carrying pollen from two orchids on the same flight. *Pterygodium pentherianum* pollen was carried on the bees' front legs, while pollen from *P. schelpei* was attached to the abdomen.

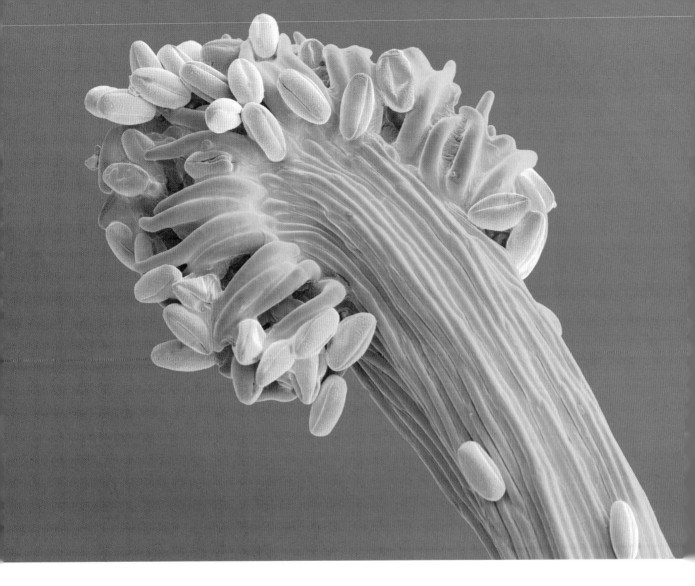

It is tempting to think that a flower can increase its chance of receiving some suitable pollen if it is open and receptive for a longer period of time. However, the flowers of some species – including those in the rockrose family (Cistaceae) – last for only a few hours, while some orchid flowers are receptive for weeks. These variable times reflect many other factors, including plant density and the number of flowers produced per plant. As stated earlier, the pollination process is influenced by an extraordinary number of variables.

The stigmatic surface plays an important role in this part of the pollen grain's journey, and is often referred to as wet or dry. Counterintuitively, the stigma's state is more important in the control of self-fertilisation (or self-incompatibility) than in pollen adhesion. Dry stigmas produce exudates that are very important in the retention of the right pollen – in other words, that from plants of the same species (conspecifics). The timing of the exudation can also be part of a strategy that aims to optimise conspecific pollen retention.

Once the pollen has landed, there are four more stages in a successful pollination: the pollen must take up water, it must be recognised, it must adhere securely to the stigma, and it must germinate. These stages do not always occur in the same order.

UNLUCKY LANDING

A coloured scanning electron micrograph (magnification: ×500) of the stigma of a gorse flower (*Ulex europaeus*) with pollen grains. Not all of these pollen grains will successfully deliver sperm to the ovule because they have landed in the wrong place.

7. Successful germination and growth towards the egg

At this point in the process of pollination there is a big difference between the gymnosperms and the angiosperms, specifically concerning access to the structure that produces the egg. In flowering plants, the eggs are produced in a sealed vessel in the ovary and access is via the stigmatic surface and down the style.

Gymnosperms

In gymnosperms, the eggs are made in a structure that is not totally enclosed and protected from the outside environment. This has a small opening in its protective layer, and this gap is generally filled with a fluid known as the pollination drop (or the pollen drop). When a pollen grain lands in the drop, one of three things will happen: it will either sink to the bottom of the drop; it will float upwards and come into contact with the female gamete-producing structure (the female gametophyte); or the drop will evaporate, and the pollen grain will be drawn onto the surface of the female. In some species of gymnosperm, such as the monkey puzzles (*Araucaria* spp.), there is no pollination drop and the pollen grain has to grow inwards from the outside of the cone. This raises the question: where does a monkey puzzle pollen grain get the water it needs for its hydration? The answer is simple. The pollen grain is not dehydrated to the same extent as other gymnosperm pollen. It is described as partially hydrated, and the rest of the water required for successful germination is probably derived from rain.

What happens next depends on which gymnosperm species you look at. In some, the pollen grain germinates, grows a tube towards the egg, stimulates the production of a fluid-filled cavity near the egg and then releases flagellated sperm into the fluid. These sperm swim to the egg and fertilise it. In other gymnosperms, the pollen grain grows a tube that is more similar to that of an angiosperm (though this may simply be a case of convergent evolution). Non-motile sperm are delivered from the end of this tube very close to the egg nucleus, with which they then fuse.

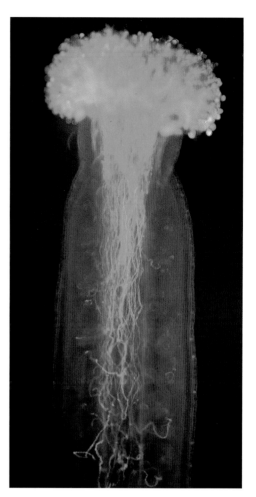

RACE TO THE OVULES

This coloured photograph of thale cress (*Arabidopsis thaliana*) shows green pollen tubes growing from the stigma, down the style and towards the ovary. Each pollen tube can fertilise only one ovule.

POLLINATION DROP

Ginkgo (*Ginkgo biloba*) is a gymnosperm and produces a drop of liquid above its egg to catch wind-dispersed pollen grains.

THE STAGES OF POLLINATION 33

Angiosperms

The meeting between a pollen grain and a stigmatic surface has been likened to the first meeting with one's partner's parents. The stigmatic surface has two questions for the pollen grain (hopefully different from those of the parents): are you from my species and are you from me?

It is clear that flowering plants can sometimes discriminate between pollen from their own species and the rest. For example, photographs taken by researchers of pollen from different viburnums (*Viburnum* spp.) on the stigma of one of the species show that while many pollen grains have germinated, only some have successfully penetrated the stigmatic surface. Some of the pollen grains have not even germinated; it is thought that these are non-viburnum pollen. How the stigma does this is not understood. It may involve the ubiquitin pathway, which as its name implies is a very widespread mechanism in plants, animals and fungi, used by cells to identify and reject cells of a different type.

Alternatively, a process referred to as unilateral incompatibility appears to use similar mechanisms to the self-incompatibility systems that recognise pollen from the same plant (see page 41). The process is described as unilateral because when pollen from a self-incompatible species is placed on the stigma of a self-compatible species, it is accepted, but when pollen from a self-compatible species is placed on the stigma of a self-incompatible species, it is rejected.

In some cases, a mechanism prevents the adhesion of non-conspecific pollen to the stigma. Investigations into this have found that sometimes pollen from the wrong plant is incapable of growing a pollination tube long enough to reach the ovary – an elegantly simple mechanism. In some crosses between closely related species, the pollen tube gets part of the way down the style, then makes a U-turn and tries to grow back up towards the stigma. In these situations, the two species are referred to as incongruent due to the evolutionary distance between them.

From this, it is clearly very reasonable to propose that plants use more than one mechanism to reach the same result. This may seem like a point of little importance, but breeders looking to overcome this reluctance of plants to cross with related species need to be able to understand all the mechanisms involved.

It is worth considering here whether the delivery of pollen from another species is a common risk faced by plants. In the garden we see bees and other flower

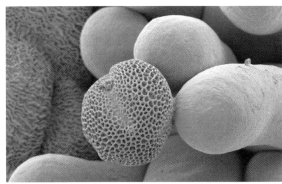

POLLEN MEETS STIGMA
Pollen adheres to the stigma (left), germinates (centre) and then the tube grows towards the style (right).

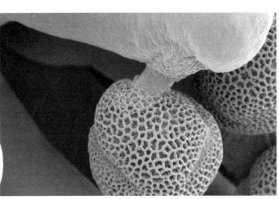

POLLEN GERMINATION
Grain is caught on stigma papillae (left, above), then the pollen tube emerges (left, below).

visitors moving between individual plants of very different species from all over the world. But does anything like this happen in the wild? Even in areas of high species richness, such as Cape St Vincent in Portugal and the Cape Peninsula region of South Africa, there is not the level of simultaneous flowering that you see in a garden. In the wild, there are often hundreds or thousands of individual plants of just tens of species. Given a small degree of pollinator loyalty, the majority of the pollen landing on a stigma will therefore be conspecific. A much bigger risk for wild plants is receiving pollen from either the same flower or from flowers on the same plant, leading to self-fertilisation.

POLLEN RECOGNITION

A scanning electron micrograph (magnification: ×12,940) of a pollen grain of *Viburnum rhytidophyllum*. The stigma of viburnums can recognise pollen from other species and reject it.

Self-fertilisation and inbreeding

Sexual reproduction exists for three main reasons. First, the reassortment of genes in the new generation facilitates new combinations that may be better suited to a changing environment. Second, sexually reproducing populations are better able to mask harmful mutations, which tend to accumulate in asexually reproducing populations. And third, sexual reproduction between unrelated individuals of the same species prevents the occurrence of inbreeding depression. This last point is particularly relevant to plants, the majority of which have bisexual flowers, where the risk of self-fertilisation is always present unless self-pollination can be prevented

Inbreeding depression is a widely observed phenomenon, whereby lineages that demonstrate a high level of inbreeding, or mating between closely related individuals, fail to thrive after a few generations and fail more severely as the generations go by. This is a particular problem in small populations where there may be no option other than to accept sperm from a close relative. The naturalist Charles Darwin (1809–1882) summed this up nicely with the phrase 'Nature abhors perpetual self-fertilization'. He knew not why, and if he had he may not have married his cousin, but he was very astute to include the word 'perpetual'. Inbreeding for one generation is unlikely to be the end of the world. In populations suffering from inbreeding depression, however, individual plants present a suite of symptoms, including reduced vigour, reduced size, fewer flowers, less pollen, fewer seeds, and smaller seeds and fruit.

Inbreeding avoidance strategies

The strategy used by gymnosperms and some woody, long-lived angiosperms to avoid inbreeding is simple yet superficially wasteful of resources. These plants go through the whole rigmarole of pollination and fertilisation, but if the embryo that forms is the result of selfing, then it simply aborts. These plants can handle the waste because they are long-lived – after all, there is always next year. Furthermore, they are large plants with many cones or flowers, and it is unlikely that all the embryos in any one year will be the result of selfing. Here, as in many aspects of plant biology, it is a question of allocation of resources. No one strategy is perfect, because life does not demand perfection; an organism just has to be good enough, which means being slightly better than the competition. As Darwin put it in the subtitle of his 1859 *On the Origin of Species*, evolution is the process of the 'preservation of favoured races in the struggle for survival'. The slightly more adequate survive, or perhaps the survivors are just slightly less mediocre.

It would be a mistake to think that each and every species follows just one strategy to avoid or reduce selfing. Some plants are dioecious and cannot self-pollinate, as the male pollen and the female ovules are

CHARLES DARWIN
The Victorian naturalist Charles Darwin was particularly fascinated with pollination, particularly that involving orchids.

on separate plants. English yew (*Taxus baccata*) is one such example, and yet in the 1990s the 250-year-old male yew tree in the University of Oxford Botanic Garden produced some female cones, for one year only. As a result, seedlings were raised from a male tree. These temporary moves from dioecious to monoecious are well recorded in flowering plants but very rare in gymnosperms.

While some plants are almost guaranteed not to self-pollinate as a result of adopting the dioecious strategy, others are almost guaranteed to self-pollinate and self-fertilise. One example of the latter group is the most studied plant in the world, thale cress (*Arabidopsis thaliana*), in which it has been calculated that outcrossing in the wild is responsible for less than 0.3 per cent of the successful seeds set. The most extreme example of deliberate self-pollination is the epiphytic orchid *Holcoglossum amesianum*, from Yunnan province in southwest China, which flowers during the dry season when there are few insects around for pollination services (see box, page 38).

YEW AND YEW

English yew (*Taxus baccata*) is a dioecious species, with separate male (top) and female (bottom) plants. It is the female trees that produce the seeds.

A case of deliberate self-pollination

In 2006, researchers from Tsinghua University in Beijing carried out research into the orchid *Holcoglossum amesianum*. They observed that, over a period of 60 days, the stalk on which the orchid's pollen is borne grows up, then down and eventually through 360 degrees to push the pollen against the stigmatic surface.

The researchers showed that all the self-pollinations resulted in successful seed-set. They also showed that the success of seed-set after manual crossing of flowers was no higher than in selfed individuals, indicating that inbreeding depression does not appear to be a problem in this species. This may mean that many generations of inbreeding have had their effects and the population is already in a genetic bottleneck.

However, the orchids do retain features of flowers commonly pollinated by insects (as are the vast majority of orchids), including a spur on one of the petals. Spurs commonly have nectaries at their far end, but the spur on the flowers of this orchid secretes no nectar, produces no odours and is also shallower than one might expect. From this, it has been suggested that the species was formerly pollinated by insects but has since discovered that self-pollination is reliable. And this was no one-off: over three years the researchers observed 1,911 flowers, all of which demonstrated the same behaviour, and they observed no insects visiting the flowers during this time.

SENSE OF SELF
A flower of the self-pollinating *Holcoglossum amesianum* orchid.

Between the extremes of English yew's dioecy and the deliberate self-pollination of *Holcoglossum amesianum* are countless intermediates showing almost every ratio of outcrossing to selfing. Perhaps the simplest and most obvious way the two strategies can be combined is by allowing any pollen from conspecific plants to land on the stigma, but relying on the pollen from another plant to grow faster and thereby win the race down the style to the ovary. In this scenario any pollen from another species should also grow more slowly than pollen from a non-selfing, conspecific plant.

The various strategies, mechanisms and behaviours adopted by plants to reduce the risk of selfing involve separating the pollen and the mature ovules in time or space, or by chemical/genetic means.

Separation in time

In cases where the parts of the flowers are separated in time, the two options are for the male structure to be produced first or the female first, respectively known as protandry and protogyny. If the separation is complete, then there will be no chance of selfing. However, complete separation is unlikely if the plant produces many flowers, because those flowers are likely to be at different stages in their development. It is only when the plant is monoecious that a complete separation is theoretically possible, and then only in the rather special case of plants like lords-and-ladies (*Arum maculatum*) and other members of the arum family (Araceae), in which all the flowers are in one structure on the plant. These plants often indulge in trapping their pollinators (see Chapter 5).

Protandry is more common than protogyny. This might be for the simple reason that the four whorls of flower parts (most commonly the sepals, petals, stamens and carpels) develop from the outside in: sepals first and carpels last. Protandry is common in some families to the point of it being a rule. As mentioned earlier, for example (see page 29), members of the daisy family have a plunger mechanism in which the pollen is produced before the stigmatic surface opens up. Protogyny is also seen in the generally monoecious genus *Euphorbia*, in which the first inflorescences produced are often made up of entirely male flowers, and only the later inflorescences have male and female flowers.

ALL IN THE TIMING

The lower flowers on the spadix of lords-and-ladies (*Arum maculatum*) are the females, separated temporarily from the males above by hairs that wither when the females are no longer receptive.

SELF-FERTILISATION AND INBREEDING 39

Separation in space

Separation of the sexes in space takes three forms. Plants can produce bisexual flowers or separate unisexual flowers. If the flowers are bisexual, then so is the plant. In this case, the amount of spatial separation is limited by the dimensions of the individual flowers and its components, but it will never be more that several centimetres. If the flowers are unisexual, there are two options. First, the unisexual flowers may be on different plants, in which case the species is dioecious and the spatial separation can be measured in metres to kilometres. Alternatively, the unisexual flowers may be on the same plant and the separation will be measured in millimetres to metres. In this latter scenario the plant is bisexual but the flowers are unisexual, and the species is monecious.

When looking at bisexual flowers and the distances that separate the anthers and stigma, it becomes clear that the relative positions can change as the flower develops. In this way the plant may be able promote outcrossing while also facilitating selfing towards the end of the flowering season. In some species the relative positions are static but vary between plants. The best-known example of this may be primulas (*Primula* spp.), which have two flower forms. In one (called the 'pin' state), the stigma is held well above the anthers, and in the other (called the 'thrum' state), the relative positions are reversed and the anthers are held above the stigma. A simple explanation of the value of this is that pollen will be deposited on the front of a pollinator that visits a pin flower. If this animal next visits another pin flower, it will simply acquire more pollen in the same place.

PIN AND THRUM

When the anthers are visible in primula flowers they are described as 'thrum' (left), whereas when the stigma is visible, they are described as 'pin' (bottom left).

MALE AND FEMALE FLOWERS

Bencomia caudata is a wind-pollinated shrub that grows on Tenerife, where the male (left) and female (right) flowers are on separate plants.

But if the next flower is a thrum, the pollen on its body will be in just the right place to be transferred to the stigma. If plants are mostly either pin or thrum, then outcrossing will be promoted. Pin and thrum is sometimes referred to as distyly, and other species have variations on this idea of moving the relative position of stigma and anthers.

Separation by chemical and/or genetic means

The final group of mechanisms to prevent selfing cannot be observed clearly without a microscope and sometimes not even in this way. This is when the pollen is rejected at the cellular and molecular level, something that has been found in at least 40 per cent of angiosperm species. There are various ways of classifying these types of self-incompatibility. The simplest is to divide them according to where the rejection happens and where the pollen fails in its mission to fertilize the ovule. This system requires 'the recognition of self' or 'self-awareness'. Along with the use of terms like this, many biologists have suggested that plants demonstrate behaviour in the same way that animals do. However, it might be better to think that animals behave while plants have strategies.

The self-incompatibility strategy is normally controlled by one gene (the S-locus) with many (usually 25–45) forms or alleles. If the incoming pollen and the receiving carpel share the same S-locus allele, the pollen will fail. This failure may happen at the stigmatic surface before it is penetrated by the pollen tube. This is quick and prevents damage to the style that could compromise future reproductive success. Alternatively, the rejection can happen as the pollen tube grows down the style. We still have a lot to discover about the molecular conversations that take place in the final stages of pollination, up to the point that the sperm fuses with the egg, by which time pollination has ended and fertilisation starts.

Advantages of self-fertilisation

Despite the many ways in which plants reduce the risk of self-pollination, many are self-compatible and tolerate self-fertilisation – as demonstrated by *Holcoglossum amesianum* (see box on page 38). In fact, more plants tolerate it than do not, because one of the most common changes in the evolutionary tree is the loss of

SHRINKING VIOLET

As the flowering season progresses, the new flowers of violets become smaller and less fragrant.

mechanisms to prevent self-fertilisation. In the sweet violet (*Viola odorata*), the first flowers of spring are relatively large and have a strong smell, and are thus promoting outcrossing. Later in the season, however, the flowers are much smaller, remain almost hidden below the leaves and self-fertilise. So why and how does self-fertilisation persist? There are four main answers to this question:

1. Self-fertilisation is reliable. As the saying goes, if you want a job done, do it yourself. Furthermore, if you are the only individual around, then selfing is the only available way to produce seeds. Plants that live on the edge of the distribution of their species or those that have colonised a new habitat, such as an island, will only survive if they can self-reproduce.

2. Self-fertilisation is cheaper than outcrossing. For plants not pollinated by animals, there is no need to produce big, showy flowers as attractants, and no need to produce an expensive reward such as nectar. Nor is there any need to produce lots of pollen (in the way that wind-pollinated plants do) because the grains do not have to travel far – millimetres rather than metres.

3. Self-fertilisation is potentially quicker than outcrossing. Thale cress is almost an obligate self-fertiliser (see page 37), which helps to explain why this little plant can become so numerous so quickly in a garden.

4. Self-fertilisation can act as a type of insurance for plants that have evolved a specialist relationship with a pollinator, in case that pollinator becomes scarce (see box).

The combination of reliable, cheaper and quicker seems irresistible, and geneticists have shown mathematically that if a plant retains the ability to both outcross and self it has a 50 per cent advantage over competitors that do not self. This is because such plants can be the male parent, the female parent or both parents of a new plant – in effect, they have three ways, rather than two, to create the next generation.

SELFING STRATEGY

Self-fertilisation is used as an insurance strategy for the New Zealand kākābeak (*Clianthus puniceus*). This plant can self but does not because the covering of its stigma physically denies access to the pollen produced by the flower's own anthers. When the pollinating bird arrives with pollen from another plant, it pushes its beak into the flower to access the copious nectar. In the process, it rubs off the stigmatic covering and in doing so exposes the stigma to pollen on its body. Should no bird visit the flower, the covering will eventually shrivel and fall off, giving the flower's own pollen the opportunity to pollinate the flower.

POLLINATION INSURANCE
The bird- and self-pollinated flowers of the kākābeak (*Clianthus puniceus*).

Chapter summary

Plant reproduction is fundamentally different from animal reproduction. Although both groups produce sperm and eggs, there are clear distinctions in the ways these are produced. In mosses and ferns, free-living plants are responsible for sperm and egg production, but in seed plants, sperm are produced and delivered in a structure called pollen. The movement of pollen from its place of production to close proximity to the place where eggs are produced is called pollination.

Pollen is small, consisting of just one to several cells. The size of a pollen grain varies greatly throughout the seed plants, and so does the colour and the patterning on the outside. However, all pollen has a coat of sporopollenin, which is one of the most resilient materials in nature.

Sexual reproduction has its advantages, but it comes at a cost. There are several mechanisms by which plants endeavour to ensure that sperm from a different plant of the same species has access to its eggs. However, there are also several ways in which plants permit self-pollination, in order to be assured of some reproduction.

Pollination can be broken down into several different stages. These include the release and presentation of the pollen, the removal and transport of the pollen to the correct destination, deposition of the pollen, and its successful germination and delivery of the sperm.

The immobile and generally non-sentient nature of plants means that they have to employ a vector to move their pollen, and in the majority of species this is an animal. Plants use a wide variety of animals as vectors, as well as wind and water. The following two chapters describe the different vectors used and the modifications to flowers that are required to accommodate these.

Chapter 2

Wind and Water

If you add together the areas under cereal crops around the world, the vast coniferous forests in the northern hemisphere, the grass-dominated savannah regions and the moorlands above the tree line, you have a significant proportion of the planet's terrestrial ecosystems. These all have one thing in common: they are dominated by plants that use wind to disperse their pollen.

Grass pollen is distributed by the wind.

Why is wind pollination so popular?

female flowers on separate plants, and the flowers vary from species to species in terms of their colour, scent, pollen coating and so on. While some species are exclusively pollinated by thrips or exclusively by wind, many are pollinated by both. It might be that these species are hedging their bets, or they may be on a commonly played out evolutionary trajectory from animal pollination to wind pollination. Each year, more species are reported as being able to exploit both animals and the wind as pollination vectors.

If you do a bit more maths and tot up the area covered by all of the world's terrestrial ecosystems, the figure comes to less than 30 per cent of the Earth's surface. Yet despite the fact that water covers the remaining 71 per cent of the surface and provides perhaps 90 per cent of the volume of the world's ecosystems, plants very rarely use that water to disperse their pollen. This is perhaps even more unexpected when you remember that sexual reproduction in plants began in the sea, with sperm swimming to an appropriate egg. Even when plants became established on land, and right up to the present day, tens of thousands of species (including the mosses and ferns, for example) still have sperm that swim through a film of water.

Why does wind appear to be so reliable and popular with seed-bearing plants, while water appears not to be? Is it *because* there are so few flowering plants, and no gymnosperms, living in water, or is it *why* there are so few flowering plants, and no gymnosperms, living in water? Or is it a bit of both?

A false dichotomy

While this chapter describes abiotic pollination (using wind and water) and the next chapter describes biotic pollination (employing animals), it is important to remember that nature abhors being put into neat boxes and is very rarely binary. When it comes to pollination, for example, there are some plants that exploit both biotic and abiotic vectors. The genus *Chamaedorea* contains about 100 species of palm from tropical America, including the parlour palm (*C. elegans*), one of the most common house plants in the world. All *Chamaedorea* species are dioecious, with male and

PARLOUR PALM
Chamaedorea elegans.

VERY SMALL AND VERY CURIOUS

The floating aquatic duckweeds (family Lemnoideae) are among the most intriguing plants in the world, as well as being the smallest angiosperms. Most of the time an entire plant consists of just one leaf that is never more than several millimetres across, and even the 'largest' plants have only four leaves and a few white roots hanging into the water below. Propagation is usually asexual and vegetative.

When a duckweed does flower, the flowers consist of two stamens that each produce about 20 pollen grains, and one carpel that produces just two seeds. The stamens are up to 2 mm (0.08 in) long and the pistil is about 1.5 mm (0.06 in) long, making them vast in proportion to the rest of the plant. But how is the pollen carried to the stigma? Perhaps wind pushes one plant against another or water currents move them? Or maybe it is transferred on the legs of tiny animals such as flies, aphids, mites and spiders? All three mechanisms appear to occur, but the latter is perhaps most likely based on the fact that the pollen is produced in small quantities, is both spiny and sticky, and has a diameter of 20 μm, putting it towards the lower limit for the pollen of wind-pollinated flowers (see page 57).

DIMINUTIVE DUCKWEED

The mini flowers of duckweed (*Wolffia* sp.) are produced by only 5-10 per cent of plants (top).

Examined up close (bottom), the flower reveals a single orange stamen (left) and red pistil (right).

Separating plants into those pollinated by abiotic or biotic vectors actually confuses matters. If we return to the moment 470 million years ago when plants emerged from the water to colonise the land, they were actually leaving an aquatic environment and entering the air. There are aquatic plants, including *Hydrostachys* species from Africa and Madagascar, that are just as well attached to soil or rocks as are oak trees and daffodils, so it was not the land that was the problem for the early colonisers, but the air. The same can be said for pollen. Most pollen is dispersed in air – sometimes on the body of an animal and sometimes on air currents, but always in air. When looked at like this, it is less surprising that there are examples of plants whose pollen is dispersed both on animals and on air currents.

FLOATING FLOWERS

Eelgrass (*Vallisneria spiralis*) has male (right) and female (left) flowers.

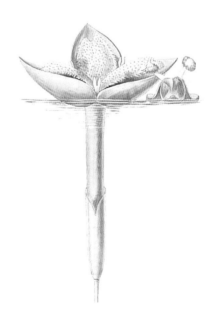

Water pollination

SEAGRASS
Pollen of submerged plants is transported by water.

Very few plant species use water as a vector to transport their pollen. Of those that do, none are gymnosperms and none are woody. The majority of plants referred to as aquatic are actually marginal aquatics, which means that they carry their photosynthetic and reproductive parts above water. In truly aquatic plants the photosynthetic parts either float on the surface, as in the water lilies (*Nymphaea* spp.) and duckweeds, or are submerged, as in the seagrasses and Canadian pondweed (*Elodea canadensis*). About a third of the true aquatic plant species are pollinated by wind, likely because their terrestrial ancestors were wind pollinated rather than their aquatic status having anything to do with it. Species that are a victim of their evolutionary history are sometimes referred to as phylogenetically constrained, and this is a common feature of evolutionary change. These wind-pollinated marginal aquatic plants are featured in the section on wind pollination later in the chapter (see page 54).

STAYING DRY
This water lily (*Nymphaea lotus*) holds its blooms above water, where they are readily accessed by flying insects. Pollination occurs entirely above the surface.

THE FIRST SUBMARINE POLLINATOR?

Work carried out on *Thalassia testudinum* in a laboratory in Mexico, published in 2016, showed that small crustaceans and bristle worms (normally 1–10mm/0.04–0.4 in long) can fulfil the four requirements of a pollinator for the species. These postulates, formulated in 1988 by Paul Cox and Bruce Knox, require that a candidate must be observed visiting both male and female floral parts (anthers and stigma, respectively), be observed carrying the pollen and be observed depositing the pollen taken from the anther onto a stigma of the same species, and that this must result in the successful formation of viable seeds in the pollinated carpel. While the laboratory experiments are not proof of animal pollination in the wild, they are a good indication that it is a possibility. For more on *T. testudinum* pollination, see page 51.

SEAGRASS
Thalassia testudinum.

Pollination by aquatic animals

While relatively few plant species use water as a vector to transport their pollen, even fewer use aquatic animals for this purpose. There is, however, at least one of these species that relies on water currents or on animals to disperse its pollen, namely the seagrass *Thalassia testudinum* (see box). Seagrasses are very important plants (see Chapter 7), and although they look like grasses they are actually members of several different botanical families.

Pollination at the water's surface

While the use of aquatic animals as vectors for pollen transport is exceptional, the use of water itself is also rare. However, some of the ways in which water is exploited are very elegant, and just as intricate as the mechanisms used on land and in the air.

The most romantic example of water transport is probably that seen in the genus *Vallisneria*. The mature male flowers of these freshwater aquatics break from their attachment to the parent plant and float up and away like a boat. The fertile stamens protrude over the edge of the boat as it floats away, propelled either by water currents or by the wind. The physician Erasmus Darwin (1731–1802; grandfather of Charles) suggested that the function of the flower's sterile stamens is to act as a sail – again, this is a blurring of the vector divisions.

The female flowers of *Vallisneria* remain attached to the parent plant and bob on the water's surface, with the highly dissected stigma protruding from between the perianth segments. They are much bigger than the male flowers and create a slight depression in the surface of the water. The tiny male then literally falls for a female, and as it drops into the depression and bumps into the female flower, the pollen on the anthers is removed by the stigma. A similar mechanism is seen in *Lagarosiphon* species.

Several aquatic species release their pollen directly onto the surface of the water, and eddies in the water or the wind moving the surface detritus propel it away from the anthers. One such species is Nuttall's pondweed (*Elodea nuttallii*), in which the female flowers sit on the water's surface and create a small depression in the meniscus (much like that formed around female flowers of *Vallisneria*). As the pollen grains reach the rim of the depression, they gently tumble into the flower and onto its stigmatic surface.

Common features and strategies

The more researchers examine the pollination of aquatic plants, the more they discover similarities in groups that do not share a recent common ancestor. This means that the strategies employed have evolved several times, a phenomenon known as convergent evolution.

One such shared feature is filamentous pollen, which can be up to 6 mm (0.25 in) long and has a propensity to clump together. The ends of some of these pollen grains have pairs of hooks to facilitate the creation of chains, which resemble strings of sausages. In other cases the pollen grains come together on the surface of the water in clumps that some people think resemble snowflakes. This clumping of the pollen is no accident, but a strongly selected strategy to increase the chance that some of it will come into contact with a stigmatic surface.

SICKLE-LEAVED CYMODOCEA
Thalassondendron ciliatum.

POURING POLLEN
Marine eelgrasses (*Zostera* spp.) shed their pollen direct into the water.

The pollen aggregations are referred to as search vehicles, because their modus operandi is reminiscent of the way in which search-and-rescue agencies increase their chance of finding victims. Some seagrasses, including common eelgrass (*Zostera marina*), employ this tactic and their stigmas have a very good success rate for 'finding' pollen. In addition, the stigmatic surfaces of the plants are highly divided and filamentous themselves, thereby increasing the chances of a pollen search vehicle hitting them. And because most common eelgrass pollen is dead within eight hours of being released, the stigma not only have to get their pollen, but they have to get it quickly.

Another way in which common eelgrass and other seagrasses increase the efficacy of their pollen capture is by having rod-shaped as opposed to spherical pollen. If pollen is spherical and is being swept downstream, it has to actually collide with a stigma to have any chance of adhesion. However, if it is rod-shaped, then it only has to get close enough to the stigma for the flow field around that structure to draw it in. This is partly because rod-shaped pollen is prone to rotate about its long axis as it makes its passage through the water.

Some aquatic plants further increase their chances of pollination success by timing the release of their pollen. One example is the seagrass *Thalassodendron ciliatum*, which grows all around the edge of the Indian Ocean from Africa to Australia. It times the release of its pollen to coincide with extremely low spring tides, which is thought to be a strategy to reduce the loss of pollen in the rougher seas that occur at other times of the year. The pollen grains also have two narrow hooks at one tip, which not only allow them to form search vehicles as described above, but also help the pollen grapple hold of the stigmatic surface.

Submarine pollination

In most of the examples illustrated above, the pollen is moved around either near or at the surface of the water. However, there is a group of aquatic plants that live entirely below the surface and so engage exclusively in submarine pollination.

As mentioned on page 49, *Thalassia testudinum* is the seagrass that potentially employs microcrustaceans and bristle worms as pollinators. It is a dioecious species, with pollen-producing male flowers on some plants and female flowers on the rest. The male plants produce about 60 flowers to every flower on the female plants. They release their pollen at night, in clumps held together by an adhesive produced in the anthers. The pollen resembles tiny frog spawn, but it does not float and so tends to congregate just above the plants on the sea floor. It is here that the pollen comes into contact with the filamentous stigma of the female flowers.

Some of the seagrasses that exhibit submarine pollination produce pollen whose density is the same as that of the surrounding water. This means that it can be found anywhere in the water column, including at the surface (apparently, the more rod-shaped grains tend to gather at the surface). Such a feature enables the species to colonise both the tidal zone and the subtidal zone.

Another common trait found in the pollen of plants that use water as a pollination vector is that it has no exine. The exine is the extremely resilient layer that characterises the pollen of terrestrial plants and that is largely made up of sporopollenin (see page 22). It is not only involved in pollen protection, but perhaps also in pollen recognition. When the soft, flexible exineless pollen comes into contact with the right stigma, it needs to be held there. In this case it appears that the adhesive that glues the pollen to the stigma is tenacious and, equally importantly, waterproof. The start of germination of exineless pollen on the stigma in the common eelgrass is marked by a molecular conversation, just as it is in terrestrial plants (see page 41). However, it is still not fully understood what is 'said' during this conversation.

LIQUID POLLEN

Male flowers of *Thalassia testudinum* release pollen and sticky mucilage into the water, the latter attracting hungry invertebrates that then distribute the pollen.

The botanical enigma that is *Ceratophyllum*

Species in the genus *Ceratophyllum* are a bit of an oddity. For starters, it has been consistently difficult to place this group accurately on the evolution tree of flowering plants, despite the wealth of data that are now available to taxonomists. It currently squats uncomfortably on a branch between the two biggest groups, the monocotyledons and eudicotyledons, and is the only genus in its family, the Ceratophyllaceae. Most botanists recognise just two species, while some think there are six species and one author has even described 30 species.

BREEDING BY BUBBLE

The filamentous foliage of hornwort (*Ceratophyllum demersum*) resembles strands of algae. This aquatic plant transfers pollen from tiny male flowers to the equally diminutive female blooms using bubbles.

Another oddity is that *Ceratophyllum* species are land plants that have recolonised the water. They are now recognised as submerged aquatic plants, and are commonly chosen for ornamental ponds in temperate gardens because they are very good oxygenators that also help to suppress blanket weed algae. They lack a root system, so all of their nutrient uptake is through their stems and leaves. The latter are so filamentous and divided that you could be forgiven for thinking that the plant in question is a member of the genus *Chara*, a group of green algae. A further strange feature of the group is that the plants store energy in a starch-based system like the one found in ferns. This is not known in any other species of flowering plant.

Peculiar pollination

The fact that *Ceratophyllum* is an enigmatic group is not really surprising. As land plants living as submerged aquatic plants, the species have had to undergo a great deal of modification, and this applies to their pollination as much as their morphology. For a start, the flowers are very odd. They are unisexual and produced on the same plant, and neither sex has any sepals or petals, although there are bracts that could be mistaken for a perianth. In the male flowers the stamens have almost no filament, and so look like just an anther. And there is still disagreement over whether the male flowers have numerous stamens or whether each stamen is a flower (similar to the situation in the unrelated and hugely diverse genus *Euphorbia*). *Ceratophyllum* pollen is what you would expect in a submerged aquatic – smooth, lacking an exine, and lacking an obvious aperture through which the pollen tube can grow. The female flowers, meanwhile, are very simple, with a single ovule and a very long style.

When the anthers are ready to dehisce, bubbles filled with gases produced by photosynthesis form within them and the pollen is released inside these. Up to 100,000 pollen grains can be produced by one flower, a scale of production that is reminiscent of wind-pollinated plants. The pollen-filled bubbles (or sometimes the whole anther) then float to the surface, where they burst. At this point the pollen grains start to sink slowly, being slightly more dense than water. Two things must then happen in order to increase the chance that the grains will hit the stigma of a female flower below. First, the female flowers reorientate themselves from an upright stance to horizontal in order to maximise exposure of grooves on the style.

TORNET HORNBLAD, CERATOPHYLLUM DEMERSUM

SEPARATE GENDERS
Hornwort flowers are very small, with separate genders, male (left) and female (right).

The second event is that the pollen grains must clump together. However, in this case the adhesion is not provided by glue from the anthers nor by small hooks, but by precocious development of the pollen tubes, which become entangled. This entangling is promoted by the fact that the pollen tubes are branched – another oddity. Because gravity brings the pollen into contact with the stigma, this means that a strong current might take the pollen downstream away from the population of plants and so reduce the chance of successful pollination; reasonably still water, or a lake, would be a more obvious habitat for these plants.

Wind pollination

Before considering the use of wind as a pollination vector by terrestrial species, it is useful to take another look at the aquatic plants that use wind pollination, which is about a third of these species. Most of these (90 per cent) have terrestrial relatives that also use wind, and most are quite orthodox in the way they exploit wind. There is, however, one example of an aquatic plant that launches its pollen into the air in a very unorthodox way.

Hydrilla verticillata is a submerged aquatic monoecious species. The male flowers are released from the plant as in *Vallisneria* species (see page 49), but at this stage they are still enclosed in their buds. When a bud reaches the surface of the water it may float for a while but then it explodes, sending a puff of pollen into the air. The female flowers open at the water's surface and their perianth effectively forms a bowl, in the bottom and centre of which are the carpels and the stigma. The pollen therefore has to fall into this chalice. On a still day self-pollination is likely, while on a windy day the pollen drifts further and may land on an unrelated flower. Of course, it could also land on the surface of the water, in which case that would be that.

FLOATING FLOWERS
The male flowers of *Hydrilla verticillata* float unattached.

Missing the target is not just a problem for *Hydrilla verticillata*, but one shared by most species whose pollen is dispersed by wind. This could be interpreted as inefficiency and might lead to the conclusion that wind pollination is simple, crude and primitive, and second best to animal pollination. However, any such thoughts are wrong, for wind pollination is as intricate and effective as animal pollination.

WIND-POLLINATED AQUATIC
The aquatic *Hydrilla verticillata* releases pollen into the air.

WIND-POWERED PLANTAIN
Broadleaf plantain (*Plantago major*) has flowers seemingly adapted for wind pollination, but also produces some nectar.

As mentioned at the start of this chapter, the idea of separating wind-pollinated plants into a group that has no relationship with animals is becoming increasingly untenable. If wind pollination is a derived state that has evolved many times from animal pollination, then we should expect to see either plants that are in the transition phase from animal pollination to wind pollination and still using both, or plants that have 'discovered' that being an ultra-generalist or supreme hedge-better is an evolutionarily stable strategy. Natural selection does not have to select the specialist, just as all-rounders still get picked to play for their country at cricket if they are good enough at batting and bowling. A botanical example here is the plantains (genus *Plantago*), which bear all the hallmarks of wind pollination and yet some produce nectar and thus are visited by pollinating bees.

Wind pollination in gymnosperms

As mentioned in Chapter 1, if you park your car under a cedar (*Cedrus* sp.) tree on a dry, still night in spring, you can expect to find it covered in pollen the next morning. It is frequently assumed that all gymnosperms are wind pollinated, because this is the mechanism used by the dominant temperate gymnosperms. Gymnosperms are also often thought of as primitive when compared with flowering plants, and so many people again commonly make the assumption that wind pollination is the primitive or ancestral state for seed plants. However, this idea soon loses credibility when we investigate the pollination of the earliest diverging extant gymnosperms, the cycads, which are either animal pollinated (generally by beetles) or both animal and wind pollinated. For reasons that have yet to be discovered, all of the gymnosperm species that are pollinated by both wind and animals are also dioecious.

Further investigation of the gymnosperm evolutionary tree reveals another oddity. Although most of the non-cycad species are wind pollinated, the very unusual collection of plants in the largely tropical genera *Gnetum*, *Welwitschia* and *Ephedra* are almost all animal pollinated and so, based on their closest living relative, they are a group that has evolved from wind-pollinated ancestors. This is uncommon. Generally speaking, wind pollination takes over from animal pollination as a result of the loss of pollinators and/or changes in growing conditions. This rare example of reversion to animal pollination in these three genera may be related to where they live, since tropical species are more likely to be insect pollinated than wind pollinated. It may also be that their extinct ancestors never completely turned their backs on animal pollination, and continued to use both wind and animals. Animal pollination in gymnosperms is covered in Chapter 3.

ON THE WIND
Cedar of Lebanon (*Cedrus libani*) releases vast quantities of pollen.

Over the years, botanists have tried to prove that wind pollination in gymnosperms is inefficient when compared with wind pollination in flowering plants, and especially inefficient when compared with animal pollination. There were several reasons why this view held sway. First, millions of pollen grains are produced for the successful production of a just a few seeds. Second, pollen from any plant species and spores from any fungus can land in the pollen drop and germinate. Third, the lack of a stigma is seen as an indication that there can be no selection or rejection of inappropriate pollen. Finally, wind is uncontrolled and erratic, meaning that the plant can do nothing to attract passing pollen grains. The fact that millions of hectares of land are dominated by gymnosperms and giga-tonnes of carbon are sequestered in these forests seems to be ignored by those who continue to dismiss wind pollination as second best. Charles Darwin considered wind pollination to be just as intricate as animal-facilitated pollination, and research carried out in the twenty-first century has backed him up on this occasion.

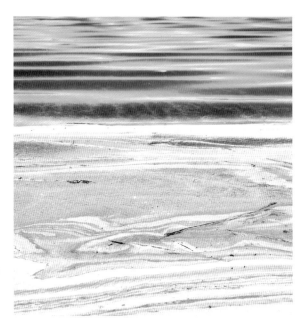

POLLEN PONDS
Conifer pollen is produced in such profusion that bodies of water quickly turn yellow under a thick film of pollen, as do cars, pavements and other structures.

AERIAL ALLERGENS
The airborne pollen of conifers such as Norway spruce (*Picea abies*) cause significant problems for allergy sufferers.

There are clearly times when wind pollination is better than animal pollination, such as when there are no suitable animals in the vicinity. But this does not address the criticisms of wind pollination listed above, such as its indiscriminate inefficiency. For example, is it true that gymnosperms do nothing to help 'attract' pollen grains towards their female cones? In order to determine whether wind pollination is inferior, it is helpful to revisit the seven stages of pollination detailed in Chapter 1: release of the pollen, presentation of the pollen, removal of the pollen by the vector, transport of the pollen, navigation of the vector, deposition of the pollen, and successful germination of the pollen grain and growth towards the egg.

The pollen of wind-pollinated gymnosperms

While plant pollen ranges in size from 4 μm to 350 μm, in gymnosperms it ranges from 20μm to 100 μm. This is because the optimum size for the pollen of wind-pollinated gymnosperms is a compromise. Smaller grains will travel further (the maximum recorded distance travelled by gymnosperm pollen is 1,300 km/930 miles), but they are also less likely to be affected by eddies around the female cone and so will not land satisfactorily. In other words, pollen can be both too big and too small to be suitable for wind pollination.

GINKGO POLLEN
Boat-shaped pollen is common in gymnosperms (×1,000).

In relation to its size, gymnosperm pollen is light for two reasons. First, it has a low water content at 5–10 per cent. And second, it often has air sacs, reducing its density compared to the pollen of flowering plants. It is tempting to think that the primary function of the air sacs is to promote buoyancy in the air, but they actually have another very important role. The female cones of many gymnosperms are heavy and therefore hang downwards, which begs the question: once the pollen grain is stuck in the pollination drop, how can it move upwards against gravity towards the ovule? The answer is that the air sacs allow it to float up to the egg. When it is close enough, the pollen grain germinates, and the events outlined in the previous chapter begin.

The shape of gymnosperm pollen varies from spherical to boat-shaped, to bowl-shaped with a central protuberance. Boat-shaped pollen is produced by cycads (which tend to be animal pollinated) and in ginkgo (*Ginkgo biloba*), so this may be the ancestral shape of gymnosperm pollen. Whether pollen shape influences the timing of pollen release, or significantly changes its aerodynamics or influences its chance of arriving at the correct female cone remains to be proved.

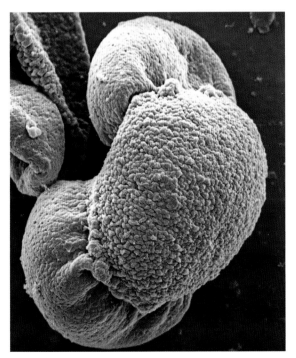

AIR SACS
Many gymnosperms equip their pollen with air sacs for buoyancy, as in this Scots pine (*Pinus sylvestris*) pollen. The two pouch-like structures keep the pollen in the air long enough for it to reach its target.

Pollen release and travel

The positioning and orientation of gymnosperm male cones appears to be determined by other features of the tree, thereby indicating that it is not random. For example, the deciduous gymnosperm dawn redwood (*Metasequoia glyptostroboides*) releases its pollen after its leaves have been shed, and deciduous larch trees (*Larix* spp.) release theirs just before the new shoots and leaves start to grow vigorously. The cones are nearly always held at right angles to the vegetative parts of the tree, possibly to maximise air flow around them. In trees with upright cones, the pollen tends to be saccate (contains air sacs), which could help it float up and away. In species where the male cones hang downwards, they tend to be on more flexible branches and the pollen does not have air sacs. Here, the movement of the branches in the wind may be enough to shake the pollen from the cone.

Once the pollen has left the male cones, its fate is largely outside its control. This journey is perilous, and death can result from many different accidents. For a start, the pollen could hit and adhere to a structure that is not a female cone of the same species. Or it may land on the ground or in water – a risk that greatly increases when it is raining. And in some cases the pollen could stay airborne for so long or travel so high (it has been detected at an altitude of 2,000 m/6,500 ft) that it suffers fatal damage from exposure to ultraviolet light or dehydration.

Returning to the question that was raised earlier of whether gymnosperms help to 'attract' pollen grains towards their female cones, it appears that features in the shape of the cones and the texture of their outer surface create eddies in the surrounding air. These eddies cause passing pollen grains to deviate towards the cones, and once they are near, they come to rest on the pollination drop.

REDWOOD POLLEN
Dawn redwood (*Metasequoia glyptostroboides*) sheds its pollen only after the needles drop.

LARCH CONES
Pollen cones of larches (*Larix* spp.) angle away from the branches.

Another alleged shortcoming of wind pollination is that there is no selection for the correct pollen or rejection of the wrong type. However, our understanding of this is also changing. In 2016, Cary Pirone-Davies and colleagues published research in support of the idea that the pollination drop is more than just a sugary syrup to receive, hydrate and feed the pollen grain. They observed differential growth of pollen grains from different species in the pollination drops of plum yews (*Cephalotaxus* spp.). The variation depended on whether the pollen grain was from the same species, from a different species in the same genus or from a different species in a different genus, and the researchers suggested that the nutrition provided by the pollination drop may suit conspecific pollen better than pollen from other species. They also suggested that the pollination drop may help to protect the otherwise exposed ovule to attack from pathogenic microbes and fungal spores.

PLUM YEW POLLEN
Cephalotaxus can recognise pollen from different species.

SEXUAL STRATEGIES OF GYMNOSPERMS

As discussed in Chapter 1 (see pages 36–7), being dioecious rules out the risk of self-pollination and self-fertilisation. It also has another advantage, in that having separate male and female plants optimises allocation of resources to being male or being female. As a result, female plants can produce larger, often fleshy, fruits. But this is biology, and so there also has to be a downside. First, if you are the only member of your species or there are no pollinators in the vicinity, then there will be no sexual reproduction. This explains why the vast majority of dioecious gymnosperms are either wind pollinated or wind and animal pollinated. And second, with dioecy only half the population can produce seeds, which can become especially problematic in small populations.

Monoecious species such as pine trees (*Pinus* spp.) face the obvious risk of self-pollination and self-fertilisation. In gymnosperms, abortion of embryos that are the result of selfing is common. However, while this may prevent inbreeding depression, it does mean that an ovule has been wasted, and prevention of self-fertilisation earlier in the process would be better. One way in which pines are thought to reduce selfing is by producing their male cones on the lower branches of the tree and female cones on the upper branches. The reasoning here is that by the time the pollen has achieved the height necessary to hit a female cone, it will have moved away from the plant that produced it.

GENDER SEGREGATION
To prevent self-pollination in pines (*Pinus* spp.), the male or pollen-bearing cones (left) appear on lower branches, while the female or seed-bearing cones (right) are held higher up.

Wind pollination in angiosperms

It is regularly and frequently stated in the pollination literature that only 12.5 per cent of flowering plants are wind pollinated. Furthermore, it is almost as regularly and frequently stated that wind pollination has evolved independently more than 65 times. This fact raises a number of questions. First, does it mean that the change from animal pollination to wind pollination is easy, and if so what are the molecular and the genetic mechanisms responsible for this change? Second, and more easy to answer, have these 65-plus transitions resulted in similar-looking plants that share particular groups of features? Third, what are the changes in growing conditions in a habitat or ecosystem that select wind pollination over animal pollination? This last question is very relevant at a time in the Earth's history when the climate is thought by many scientists to be changing faster than ever before.

The pollen of wind-pollinated angiosperms

The first thing that strikes you about the pollen of wind-pollinated flowering plants is just how much there is of it. The number of pollen grains produced by wind-pollinated plants can be hundreds or thousands of times more than that by produced animal-pollinated plants for the same number of seeds set. As stated before, pollen ranges in size from 4 μm to 350 μm, and while gymnosperm pollen ranges from 20 μm to 100 μm, the pollen of wind-pollinated flowering plants ranges from 20 μm to 40 μm. The minimum and maximum sizes are determined by the need to be large enough to stand a good chance of hitting a stigma, but not so large that the pollen grain plummets to the ground too early. The largest pollen of wind-pollinated flowering plants tends to be smaller than that of gymnosperms because, unlike the latter, it rarely contains air sacs.

Due to differences in its pollenkitt, the pollen of wind-pollinated angiosperms is generally dry rather than sticky (as is the case in animal-pollinated species). The lack of a sticky layer prevents clumping of the pollen and therefore improves its ability to travel a reasonable distance to another plant. (This is in contrast to the pollen of water-pollinated plants, which is more likely to hit a stigma if it is stuck together in clumps – see page 50). In addition, the exine is smooth rather than sculptured, making it more aerodynamic so that it travels further, and it often has one or no preordained apertures for the pollen tube, which may be an adaptation to reduce water loss during transit.

GRASS POLLEN

Wind-pollinated angiosperms such as grasses shed pollen in great quantity and are a major contributor to human allergies.

GRASS FLOWERS

With no need to attract pollinators, golden oats (*Stipa gigantea*) have plain flowers.

The flowers of wind-pollinated angiosperms

The pollen of wind-pollinated angiosperms is 'presented' to the wind on exposed anthers. In the grass family (Poaceae), the anthers are on long, flexible filaments. They open when the pollen is ready to be released – the whole anther splits and twists, allowing the pollen to tumble out. This is more likely to happen in dry weather and when there is wind, as both these factors will contribute to the desiccation of the anther prior to splitting.

Many non-grass wind-pollinated plants have separate male and female flowers, which may be on the same plant, as in the pedunculate oak (*Quercus robur*) and common hazel (*Corylus avellana*), or on different plants, as in *Garrya elliptica* and *Bencomia caudata* (see page 40). In all of these situations the male flowers are grouped together in catkins. These pendent tassel-like structures consist of many small flowers, which can release a very large number of pollen grains. In some cases, as in birches (*Betula* spp.), the catkins have small flaps that open only when the wind is strong enough to carry the pollen a reasonable distance from its parent plant. In dioecious species, the female flowers are often also held in catkins.

WIND-POLLINATED TREES

Many deciduous trees in temperate areas utilise wind pollination. Their height allows the pollen to travel some distance and, as the flowers often appear before the foliage, the leaves do not impede air flow. Both pedunculate oak (*Quercus robur*; top) and common hazel (*Corylus avellana*; bottom) have catkin-like male inflorescences and solitary female flowers.

In monoecious wind-pollinated species such as the oaks, birches and hazels, the female flowers are generally smaller than the male flowers (much smaller in the case of the hazels). The reason for this difference is not known, but it might have something to do with the optimum proportion of energy the plant should allocate to its male and female function if it is bisexual. Monoecious species face the perennial problem of possible inbreeding (see page 36), as do the grasses, whose flowers are bisexual (although note that maize, *Zea mays*, is a rare example of a monoecious grass). A simple way plants reduce this risk is by separating the male phase from the female temporally, and in the majority of these cases the male function comes first.

Because wind is blind, the flowers of wind-pollinated plants are noticeably dull in terms of colour and have a much-reduced perianth. Furthermore, the wind has no olfactory sense and so the flowers do not produce any odours. And because the wind will turn up irrespective of whether there are plants to pollinate, there is no need for the flowers to supply any type of inducement or reward such as nectar.

Pollen arrival

Once the pollen has matured and been released onto the wind, its target is the stigma of a unrelated plant in the same species. It is no surprise that the female flowers of wind-pollinated angiosperms have a very divided stigma, which may be feather-like. The stigma and style are normally attached to an ovary that will produce one seed. This makes the pollen-to-ovule ratio even higher than that found in animal-pollinated flowers.

The pollen–stigma interaction is critically important because the stigma has to hold onto the correct pollen and recognise it, and it may have to do this very quickly. This is achieved through a variety of mechanisms, including charged pollen grains that are drawn towards a stigma. There is also good evidence that in some species, including jojoba (*Simmondsia chinensis*), the flowers are held relative to the leaves in such way as to create eddies that make the pollen more likely to hit the stigma. Unlike the evergreen jojoba, many wind-pollinated angiosperm trees are deciduous, and they release their pollen before the plants are in full leaf (this strategy is also seen in the deciduous gymnosperms).

MAIZE FEMALE FLOWERS
The silky stigmas collect pollen from the air.

MAIZE MALE FLOWERS
Numerous stamens release pollen in great quantity.

JOJOBA FLOWERS
In jojoba (*Simmondsia chinensis*), leaves divert pollen towards female flowers.

A habitat fit for wind-pollinated plants

Plants with wind-pollinated flowers are found only in suitable habitats, but what are these? If you look at the distribution of wind-pollinated plants around the world, you will discover that they are most commonly found in temperate regions and/or open habitats. You are unlikely to find them in tropical woodland, because there is simply not enough wind in this dense vegetation and there are plenty of animals with the qualifications needed to be a successful pollinator. However, there is a critically important point to note here. In the absence of self-compatibility, the distance from one wind-pollinated plant to another must not be too great and the path between them must be kept clear. Because there are so many different species in tropical woodland it is unlikely that there will be a straight line between two plants of the same species. Temperate woodland is depauperate in comparison, and so wind pollination is a viable option here.

WINDY WOODLAND
Temperate prairies and woodlands are rich in wind-pollinated plant species.

Outside of tropical woodlands, where habitats are less dense and the seasons become more pronounced, it is reasonable to propose that some areas face a shortage of pollinators, either seasonally or completely. Furthermore, seasonal changes can provide a reliable signal for plants to enter the next phase in their life history. This can then lead to all the individuals in a species flowering reasonably simultaneously, thereby facilitating cross-pollination and outcrossing.

It has been suggested that angiosperms may have experienced a shortage of animal pollinators soon after their emergence in the Cretaceous period, 145–66 million years ago. In these conditions, plants that could use wind as well as an animal as a pollination vector, such as the modern-day cycads (see page 55), would have succeeded better than those that could not. Of course, there are other possible outcomes in this scenario. The plants may have just failed and become locally extinct. Or they might have adopted a self-pollinating strategy that got them through the temporary difficulty. Alternatively, they might have been attractive to another animal species.

Another reason why wind pollination is unlikely to thrive in those tropical habitats where rainfall is a regular feature is that rain is very bad news for airborne pollen. Temperate forest, on the other hand, such as the magnificent woodlands of the Olympic Peninsula in Washington state, USA, have a sufficiently long period of dry conditions for the gymnosperms to succeed. The trees in the Atlantic Northeast region of North America are also pollinated by wind, but these forests are deciduous and mostly flowering species. From this, you might hypothesise that tree genera found in both temperate and tropical regions display different pollination strategies, and you would be right. Temperate oaks (*Quercus* spp.), for example, are wind pollinated, while tropical oaks in the same genus are pollinated by animals.

One final habitat where wind pollination could have a strong selective advantage is where wildfire is a regular and frequent event. In such areas it may be difficult for populations of pollinators to survive or, if they can survive, to exist at sufficient densities. Annual fires are a common feature of tropical savannahs, and here the dominant plants are grasses – all of them wind pollinated.

PERFECT HABITATS

In places where pollinators are in short supply, such as frequently burned grasslands (above) and seasonal woodland (left), wind pollination is common.

Chapter summary

Wind and water are important vectors for pollen. Wind-pollinated plants share a suite of ecological requirements and morphological features, and water-pollinated plants show more floral diversity than wind-pollinated plants, but there are a few features common to both groups as a result of using a non-animal vector. In both wind- and water-pollinated plants the stigmatic surface is highly divided, long and exposed in order to increase the chance of a pollen grain hitting it. And in both groups there is an almost ubiquitous incidence of unisexual flowers, which are often on separate plants (dioecy). However, there are differences between these two groups in every other aspect of the seven stages of pollination.

Wind and water pollination are not simple, and they are not the ancestral condition; both are derived strategies. This means that the ancestors of these plants were pollinated by animals. The next chapter considers those animals and the factors that led to their association with plants as pollinators as far back as perhaps 400 million years.

Chapter 3

Animals

Scientists have calculated that at least 315,000 plant species are pollinated by animals and that nearly 350,000 animal species are involved in pollination. If these figures are correct, then it means that up to 10 per cent of terrestrial animals derive some, if not all, of their nutrition from plants in the form of nectar or pollen, or both, in return for their services as pollinators. And that does not include the unknown number of animals that visit flowers for food but do not pollinate the plants. What we do know is that a very wide range of animals is pollinating flowers all over the world, as this chapter will explain.

Australian banded honeyeater (*Cissomela pectoralis*) on Eucalyptus flowers.

Pollinator job description

If 350,000 different animal species are pollinators, what do they all have in common and what does it take to be a pollinator? If you were the recruitment consultant for a plant species looking for a pollinator, the job description might look something like the following.

The key responsibilities of the pollinator

The pollinator is responsible for the safe and timely transport of pollen from the male reproductive structures on one plant to the female reproductive structures on another plant of the same species, or if another plant is not available, to the female reproductive structures on the same plant.

The duties of the pollinator

- To visit flowers or cones when they are shedding pollen;
- to visit female cones when receptive or flowers when their stigmatic surface is receptive to pollen;
- to return each year to perform the same duties, or to send offspring that have been prepared appropriately for the task;
- to carry the pollen safely and efficiently; and
- to take the reward from the flower but nothing else.

Optional duties might include:
- telling conspecific animals where the plants may be found; and
- good botanical identification skills.

Personal specifications of the pollinator

- Share a similar habitat to the plant;
- be reliable – able to visit at the same time each year when required;
- be part of a community of similar animals that is numerous enough to visit all the flowers in the plant population;
- be physically fit enough to travel to visit many plants in a day;
- be the correct size to visit the flower to remove or deposit the pollen without damaging the flower; and
- be omnivorous, vegetarian or vegan (no carnivores).

Payment

- Food in the form of nectar, pollen or starch;
- a mate;
- warm shelter for the night;
- somewhere to lay eggs and a nursery for the young (females only); and/or
- pseudo-copulation (male bees and wasps only).

Additional information

Be aware that some employers will make false promises and put out hoax signals.

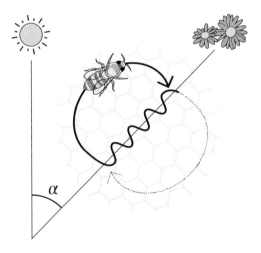

WAGGLE DANCE
Honeybees (*Apis mellifera*) can indicate the direction and distance of a food source to hive mates.

BATS
These night-time nectar feeders visit a wide range of flowers. This is the Asian cave nectar bat (*Eonycteris spelaea*).

BIRDS
Often travelling long distances, birds make efficient pollinators. This rainbow lorikeet (*Trichoglossus moluccanus*) has a brush-tipped tongue to lap nectar.

BUTTERFLIES
Most butterflies, like this swallowtail (*Papilio machaon*), are good pollinators partly helped by their generally excellent colour vision.

POLLEN BEETLES
Common pollen beetles (*Brassicogethes aeneus*) can cause significant damage to flowers, although they may also pollinate some species.

This anthropomorphic interpretation of the requirements of a suitable animal pollinator enables us to identify potential candidates and eliminate others. For example, some birds will be too big for the role and spiders are unlikely to be successful unless they are predators of the actual pollinators (in the way that crab spiders prey upon honeybees) and accidentally remove pollen and take it to another flower.

The range of animals that have been inveigled into pollinating plants is extraordinary, and a testament to the diversity of nature and the ingenuity of evolution by natural selection. It is no surprise that Charles Darwin drew so much supporting evidence for his ideas on natural selection from pollination relationships. The range of sizes of those involved is several orders of magnitude, from vertebrates such as possums, lizards and parrots, to tiny invertebrates such as ants, thrips and biting midges (the latter pollinate the flowers of the cacao tree, *Theobroma cacao*, the source of chocolate).

The proximity of the number of plant species pollinated by animals (315,000) to the number of animals that pollinate plants (350,000) should not be interpreted as significant in any way, for several reasons. While we know that some plant species are pollinated by just one animal species, these represent the minority and many have multiple pollinators, just as many animals pollinate many plants. Also, our catalogue of the world's animals is woefully incomplete, especially when it comes to invertebrates. As this group accounts for about 99.6 per cent of known pollinators, there must be many that have yet to be recorded. This lack of a complete world species catalogue is sometimes referred to as the taxonomic impediment, and is discussed further in Chapter 8.

Vertebrate pollinators

Vertebrate pollinators can be divided either into flying and non-flying species, or into reptiles, birds and mammals. There do not appear to be any records of pollination by non-flying birds, and so birds and bats are the flying vertebrates, and lizards and a mixed bag of mammals are the non-flying vertebrates.

Lizards

It is difficult to make a case proving that lizard pollinators are a globally significant group. For a start, nearly all are found on islands such as New Zealand and Crete, or island groups such as the Balearics, Macaronesia and those in the Indian Ocean (for more on this, see Chapter 6). In addition, they have been overlooked as pollinators because it was originally assumed that they are all carnivorous. While only 1 per cent of lizards are truly herbivorous (deriving the majority of their food from plants), it now appears that more species than previously thought are, in fact, omnivores – yet another example of the blurred lines in the living world.

UNLIKELY POLLINATOR
While some lizards pollinate plants, this Madeira wall lizard (*Lacerta dugesii*) is also likely to feed on a pollinating insect.

A REPTILIAN FACILITATOR

If you are looking for unusual evolutionary developments, oceanic islands are a very good place to start, and Tasmania is home to one of the strangest. The snow skink (*Niveoscincus microlepidotus*) and the honey bush (*Richea scoparia*) are both found near the summit of Mt Wellington, overlooking Hobart. The flowers of the honey bush are protected by calyptra, which are actually fused petals. The skink tears off the calyptra to get at the nectar, and in the process also makes the flower accessible to wasps, flies, bees and other general pollinators. Furthermore, even if the unopened flower was able to self-pollinate, the resulting fruit cannot open to release the seeds unless the calyptra has been removed. In this case, therefore, the skink is not the actual pollinator, but it is an essential facilitator of pollination.

SNOW SKINK
Niveoscincus microlepidotus is not a pollinator, but rips open the flower buds of *Richea scoparia*.

BUFF-TAILED BUMBLEBEE
Pollinating bumblebees like *Bombus terrestris* can access blooms only once they have been opened by a snow skink.

Lizards are different from the other vertebrate pollinators in that they are ectothermic, which means that they rely on external heat sources to regulate their body temperature. This is relevant here because, unlike endotherms (which generate their body heat internally), they do not have to eat regularly. Plants do not flower incessantly, and so the lizards can use their nectar and pollen as and when it is available. One of the desirable characteristics of pollinators is loyalty to a species, and observations of gecko behaviour in New Zealand have shown that different species of gecko remain very loyal to a species of flowering plant when its nectar is on offer.

However, there is one way in which lizards do not seem to fit the pollinator job description, and that is size. Generally speaking, an adoption of herbivory is accompanied by an increase in body size, and the minimum size for a herbivorous lizard is 100 g (3.5 oz). Therefore, pollinating lizards are more likely to be omnivores and small rather than pure herbivores. This is supported by the evidence, which shows that lizards as small as 25 g (0.9 oz) are pollinators. The size is important, because a large lizard crashing around on a plant looking for nectar could inflict a lot of damage in the process.

MONKEY POLLINATORS

Several primates pollinate plants; for greater efficiency, they may visit each bloom along a route, returning to the first flower only after its nectary has refilled. This toque macaque (*Macaca sinica*) is more likely eating the flowers!

Non-flying mammals

The contribution that non-flying mammals make to pollination was not really acknowledged until the 1930s, and the first work on the subject was published in German, which resulted in a gap of another 40 years until scientific reports started appearing in English-language journals. One problem that the scientists faced was proving that these animals had actually carried out the work. It is one thing to recruit pollinators, but until they have been in the post for at least a year it is impossible to know whether they have been fulfilling the role according to the job description. At that first-year review there are three things the employee will need to prove: first, that the pollinator has been visiting the right flowers regularly; second, that it has been carrying significant amounts of pollen; and third, that its work has resulted in the production of viable seeds.

In the case of mammalian pollinators, finding supporting evidence is a bit tricky since many are nocturnal and a lot of their work is undertaken at night. In addition, direct observations are required because researchers have to see the movement of pollinators between flowers on plants of the same species. From these observations, it appears that some of the plants pollinated by mammals are also visited by birds.

At least 108 species of non-flying mammals have now been implicated in pollination, with marsupials, rodents, primates, tree shrews, elephant shrews and carnivorans among their number. Originally, all these were confined to parts of the former southern supercontinent of Gondwana, including South America, Africa and Australia. At least 85 plant species are pollinated by non-flying mammals and most, if not all, are visited by more than one pollinator and each pollinator visits more than one plant species. This means that the plants supply nectar over a longer period than if they had a one-to-one relationship with a pollinator.

One of the issues surrounding the suitability of mammals as pollinators is how far they travel and whether this is far enough to service a significant proportion of the population of a plant species. Monkeys in Africa have been observed trap-lining the flowers of *Daniellia pynaertii*, a large tree in the bean family. Trap-lining is a behaviour shown by many pollinators (not just mammalian), which sees them systematically working their way through the flowers on a plant and through the plants of the same species in an area. They will also visit this group of plants on several occasions.

Because mammals are endotherms, another question concerning mammalian pollinators is what they eat when the plants are not in flower. Fruits seem to be one obvious answer, and in India at least this has been proved. Here, *Cullenia exarillata* is pollinated by non-flying mammals that are well known as fruit eaters, and the tree's flowers seem to fill a gap in the year when fruits are not available. The structure of the flower seems to support this idea, because the nectary can be accessed only by biting it off with the sepals. This sweet-tasting reward is thought to provide similar nutrition to the fruits the pollinators otherwise feed on. (For more on pollinator rewards, see Chapter 5.)

Assessing how important and effective non-flying mammals are for successful seed-set compared with the birds that visit the same flowers involves experiments in which one of the guilds is excluded. There are few data on this at present owing to the difficulty of excluding the pollinators, but where such experiments have been carried out successfully (on species of *Banksia* in Australia and *Protea* in South Africa), the results show that the mammals are at least as good as the birds – and in one case much better. However, it has been suggested that non-flying mammals are important only where flower-visiting bats are rare or absent.

DORMOUSE POLLINATOR?

Although many mammals are observed feeding on flowers, it is not always easy to discover whether they pollinate them. Most are simply opportunistic, consuming the flowers, nectar or attending insects.

TRAVELLER'S TREE

The robust green bracts around the flowers are more than capable of withstanding the impact of hungry lemurs.

THE LEMUR AND THE TRAVELLER'S TREE

Some plants and their animal pollinators are closely linked. One such relationship is that between two iconic endemic species from Madagascar, the traveller's tree (*Ravenala madagascariensis*) and the black-and-white ruffed lemur (*Varecia variegata*). The flowers of the tree are robust enough to cope with the handling they receive, and at the same time the lemur is very careful not to damage them. The nectar is a significant component (22 per cent) of the lemur's diet, which also includes fruit and leaves. Likewise, the lemur is not the tree's only pollinator. For these reasons and also the fact that the tree can self-pollinate, the lemur needs the tree more than the tree needs the lemur.

Bats

Bats account for the majority of mammalian pollinators – 236 species out of a total of 344. The simple distinction of being able to fly is enough to put them well ahead in the queue for candidates, in front of the non-flying species. This view of bats as effective pollinators is supported by the fact that bat pollination appears to have evolved independently more than 60 times across the plant evolutionary tree. In addition, bat pollinators do not appear to show any taxonomic bias for particular groups of plants as long as they can fulfil a few simple criteria, such as the production of a lot of nectar.

There are approximately 1,200 bat species worldwide, so almost 20 per cent of them pollinate flowers. These 236 species are on record as pollinating 528 species of flowering plants. This ratio differs from that described above for non-mammalian pollinators and tells us that, on average, bats do not form one-to-one relationships with flowering plant species. In fact, there are many examples of groups of plant species that are pollinated by groups of bat species. This is no surprise, however, because generally bats are active year round and, unlike some pollinators such as hummingbirds, cannot enter a temporary state of torpor when the going gets tough. Hence, pollinating bats have a rota of plants that they pollinate through the year and that they rely on as sources of food.

Bats fulfil many of the criteria listed on the pollinator job description. They can carry large loads of pollen, and with careful, accurate deposition of pollen from the anthers of the flowers they can carry that from a few different species at the same time. They can also carry pollen a long way – it is not unusual for a bat to fly 50 km (30 miles) from its roost site, although distances of half that are more common. This means that bats are very suitable pollinators where the flowering plants are at a low density and widely dispersed. It also means that they are very good at promoting outcrossing between groups of plants that are large distances apart, whether these groups are at a high or low density, thereby giving the species' gene pool a good stir. This has been proven through observations of seeds with different male parents being harvested from the same bat-pollinated plant. In fact, it has been shown genetically that bats are better at facilitating gene flow and reducing selfing than insects. This promotion of outcrossing is actually also true of birds, which explains why there are very few vertebrate-pollinated dioecious plants.

The sensory and cognitive powers of bats are a cut above those of some invertebrate pollinators. Despite rumours to the contrary, some species have good visual skills and some even see in the ultraviolet spectrum, which means that they can detect patterns on petals that are invisible to humans. They also have a very keen sense of smell and are not put off by odours that we find unattractive. In addition, the New World species of bat pollinators use echolocation, making them extremely accurate at homing in on individual flowers once they have been drawn in most of the way by scent.

Bats have excellent spatial awareness and spatial memory, allowing them to return to the same plants over a series of nights. They demonstrate trap-lining, foraging to optimise the volume of nectar they can collect. They can also communicate with their roost-mates and help them find the good flower sites. One feature of bat behaviour that may reduce the gene flow of the plants they pollinate is when males guard their territory. In these cases it is quite possible that long-distance pollen distribution may be compromised by aggressive males.

A COAT OF POLLEN

Flowers that attract bats typically open at night and have robust petals and other structures, as these pollinators are often large and may clamber across the flower cluster.

HOVERING BATS
Bats native to the New World hover over flowers, as with this southern long-nosed bat (*Leptonycteris curasoae*) feeding on *Agave* flowers.

Not everything about bats makes them perfect pollinators. First and foremost, they are high maintenance and consume a lot of energy when flying. Bats work on a daily energy budget of 40–50 kJ (9.5–12 kcal), which is many tens of times greater than that of an insect. The New World bats weigh 15 g (0.5 oz) on average and they hover when feeding, while the Old World species weigh 38 g (1.3 oz) on average and they land, which can damage the plant if it is not robust enough. In all cases the average visit to a flower is brief, rarely lasting more than two seconds. This means that the bat has to grab as much food as possible very quickly. In order to expedite this, the New World bats have longer snouts and longer tongues than the average bat (in fact, the longest mammalian tongue, relative to the size of the animal, is found in the mouth of a bat), which is believed to improve their nectar-collecting capacity. A further modification is the presence of papillae the size of hairs on the tip of the tongue.

PERCHING BATS
Pollinator bats in the Old World, such as this cave nectar bat (*Eonycteris spelaea*), tend to land on the flowers they feed from. This banana plant (*Musa* sp.) therefore has thick, leathery bracts and robust flowers. Abundant nectar is another requirement to attract hungry bats.

Birds

The recruitment of birds as pollinators, mostly as replacements for insects, has happened dozens of times, just as it has with bats. Some 1,089 species of birds have been recorded as effective pollinators, or about 10 per cent of all bird species. Why this percentage differs from that in bats is unknown, although it is interesting to note that bats are perhaps more needy than birds. Bat-pollinated flowers on some cacti produce 8–20 times the nectar calories per flower as that found in the flowers of species pollinated by hummingbirds. It has also been suggested that bird-pollinated flowers have to invest in extra defences against nectar robbers, which take the reward without doing the work.

Deciding whether a bird is actually fulfilling the role of pollinator is not always easy to see at first glance, because some birds visit flowers to prey on the insects that are actually doing the job, just as crab spiders do (see page 117). The birds that are proven to be bona fide pollinators mostly come from four groups: the hummingbirds, from North, Central and South America; the sunbirds, from sub-Saharan Africa, Arabia, Southeast Asia and Australia; and the honeyeaters from Australia, New Zealand and the Pacific islands to Hawai'i. In addition, lorikeets, flowerpeckers and orioles all visit flowers and can pollinate them.

Generally speaking, bird pollination is more common in regions where there are no distinct seasons and where plant growth and flowering occur at all times of the year. These are most commonly tropical and subtropical areas of scrub, open woodland, and riverbanks. In more seasonal regions, the bird pollinators are more likely to be migrant species. There is a classic biological paradox here, because it is assumed that many plants have adopted bird pollination to circumvent the problem of insect shortages, and yet some of the birds that engage in pollination (such as the hummingbirds) also eat insects, and indeed are often seen eating them off flowers.

Some of the traits that make bats suitable as pollinators also apply to birds. Just like bats, birds can carry large loads of pollen a long way, thereby connecting dispersed plant populations or individuals. In both cases the gene flow is enhanced, and the gene pool is added to and potentially improved. Again like bats, birds have exceptionally good visual skills, and many have been shown to have an excellent memory when it comes to finding previously visited plants and populations. Birds are also less likely to be put off by bad weather in the way that bees are, and there is good evidence that birds often feature as pollinators where insects fail to thrive or even exist, such as on oceanic islands.

SUNBIRDS
The malachite sunbird (*Nectarinia famosa*) is a common bird pollinator in Africa and Asia; this individual is feeding on *Aloe*.

OPPOSITE: HUMMINGBIRDS
Restricted to the Americas, this hummingbird has prickly-pear cactus (*Opuntia* sp.) pollen on its beak.

HONEYEATERS
Primarily from Australia, honeyeaters include the New Zealand tūī (*Prosthemadera novaeseelandiae*), seen here feeding on flax (*Phormium* sp.).

BIRD POLLINATION IN EUROPE

It appears that bird pollination is currently almost absent from Europe, as well as from most of Asia north of the Himalayas. There are fossils of hummingbirds in Europe, suggesting that bird pollination was once found there, but this is only circumstantial evidence. In 2005, a published report recorded that in Spain *Anagyris foetida* is pollinated by chiffchaffs (*Phylloscopus collybita*), Eurasian blackcaps (*Sylvia atricapilla*) and Sardinian warblers (*S. melanocephala*). Some of the individuals in the first two species may well be migratory birds from northern Europe. This is an important point because, as with the bats, birds need to be fed year-round and so a winter food supply such as the nectar from this plant is very important.

Hummingbirds

The hummingbirds (family Trochilidae) are by far the most studied of the bird pollinators and account for about a third of the total. They are mostly tropical and subtropical species, although some migrate to temperate USA in the summer, with some resident in western States. Hummingbirds do not feed exclusively on nectar, with many species also eating insects. For this reason their beaks are adapted for both catching insects and probing nectar-rich flowers, and they also have a very flexible jaw.

The hummingbirds differ from the other three major groups of bird pollinators in that they are not passerines and so are not perching birds, which means that they hover when collecting nectar. Furthermore, they are not gregarious to the same extent as passerines and this limits the size of the plants they pollinate – they do not pollinate large trees, for example, as these produce many flowers at a time and so tend to attract flocks of birds.

Broadly speaking, there are two groups of hummingbirds – hermits (subfamily Phaethornithinae) and non-hermits (subfamily Trochilinae). The hermit hummingbirds are non-territorial and are trap-liners, which means that they are good at promoting cross-pollination between plants. The non-hermit

HOVERING HUMMINGBIRDS
Unique among bird pollinators, hummingbirds such as these long-tailed sylphs (*Aglaiocercus kingii*) can hover in front of flowers.

HOT COLOURS
Bird pollinators, including this rufous-breasted hermit (*Glaucis hirsutus*), are often attracted to flowers with warm hues, especially red, orange and yellow.

hummingbirds are territorial, and males will often defend a clump of 'their' plants. This might guarantee gene flow between the plants in that clump, but it will reduce genetic diversity being brought in from elsewhere by these males.

Sunbirds

Among the sunbirds (family Nectariniidae), 124 species have been recorded as pollinators. One of the most charismatic groups of plants to be pollinated by sunbirds are the southern African bird-of-paradise flowers, and in particular the giant white bird of paradise (*Strelitzia nicolai*), which produces very large white-and-blue flowers year-round. It is pollinated by four species of sunbird, which compete for the feast provided by the flowers; of these, the olive sunbird (*Cyanomitra olivacea*) is generally the winner in any squabbles. The individual birds have to be heavy enough to open up the flaps covering the anthers, so that the pollen is attached to their feet. The pollen comes off in clumps because it is held together in sticky threads. The birds spend much more of each visit to a flower standing on the anthers than they do standing on the stigma, and yet this is a very effective pollination system.

STRONG SUNBIRD
The olive sunbird (*Cyanomitra olivacea*) is robust enough to part the flaps covering the pollen in a bird-of-paradise flower.

GIANT WHITE BIRD-OF-PARADISE FLOWER
The blue petals of *Strelitzia nicolai* form a perch.

Honeyeaters

The honeyeaters (family Meliphagidae) are an important group of pollinators in Australia, where more than 1,000 plant species are pollinated by birds. Whereas hummingbirds often form close relationships with a few specific species, most of the birds in Australia visit a wide range of flowers and most plants are visited by a wide range of birds. Not only that, but some eucalypts (*Eucalyptus* spp.) are pollinated by birds in the early morning, followed by bees, flies and beetles through the day, and then nocturnal mammals and moths at night. The birds involved often have brush tongues, which helps them to collect the nectar. The pollen is usually gathered passively during feeding and can be placed almost anywhere on the front end of the bird (but apparently not the feet).

One explanation for why bird pollination is so prevalent in Australia is that the soils are generally poor in nutrients. Using pollen as a reward is expensive for the plant in terms of nitrogen supplies in particular, but to make sugars for lacing nectar it simply needs some carbon dioxide, water and sunlight, the last of which is in abundant supply in many parts of Australia.

HUNGRY HONEYEATERS

Honeyeaters browse a wide range of plants; this blue-faced honeyeater (*Entomyzon cyanotis*) is feeding on grevillea.

Invertebrate pollinators

In 1988, Robert May, Baron May of Oxford and former president of the Royal Society, wrote that 'by far the most abundant category of recorded living species is terrestrial insects. To a rough approximation and setting aside vertebrate chauvinism, it can be said that essentially all organisms are insects.' This is a provocative statement in a passionate and emotional scientific paper, but it does make the important point that there are an awful lot of invertebrates.

The exact number of invertebrate species worldwide is unknown, but a widely acknowledged figure for the named species is 1.25 million, of which 352,000 (28 per cent) are recorded as flower visitors. Remembering that 20 per cent of bat species and 10 per cent of bird species are pollinators, it would appear that insects are more suited to the work. Of course, not all invertebrates that visit flowers are pollinators. For example, spiders are there to prey on other visitors, some of which will be pollinators and some of which will be herbivores that would otherwise damage the plants and their flowers.

Plants are clearly essential for the terrestrial diversity of insects, but insects have become almost as important for the diversity of plants, because 99.6 per cent of pollinator species are invertebrates and the vast majority (99.9 per cent) of these are insects. And as pointed out at the start of the book, more than 87.5 per cent of plants are pollinated by animals. So what is it about insects that makes them such good pollinators?

A backwards way of investigating this question is to ask what it is about the groups of insects that do not play a part in pollination that makes them unsuitable for the role. There are many different answers to this. Some insects, such as the dragonflies, damselflies and mantises, are carnivorous and so more interested in the pollinators than the pollinatees. Indeed, the snakeflies are such aggressive carnivores that pollen – presumably originally attached to their prey – has been found in their stomach.

HOVERFLIES
Insects are the most common animal pollinators of plants and encompass great diversity; this hoverfly is feeding on *Anemone*.

Some invertebrates are aquatic, although this does not rule them out as pollinators – as seen in Chapter 2 in the example of crustaceans and marine worms theoretically pollinating seagrasses. Some insects do not feed in their winged adult stage and so are not interested in the free lunch offered by flowers. Some have very thin exoskeletons and need to stay in sheltered habitats with high humidity, and are therefore too exposed on flowers. Some cannot fly, which is a serious drawback when it comes to pollination. And some are internal parasites of other insects.

The majority of insects that visit flowers (98.6 per cent) belong to just four groups: moths and butterflies (Lepidoptera); beetles (Coleoptera); bees, ants and wasps (Hymenoptera); and true flies (Diptera). So do these groups share characteristics that make them particularly suited to the role of pollinator? One suggestion is that, in order to eat pollen, insects do not need any special adaptations beyond those for general herbivory. This means that insects were 'pollinator ready', or exapted (see page 12), when pollen first evolved hundreds of millions of years ago. Specialised mouthparts needed for sucking up nectar came later.

The fact that most insects in the groups named above can fly is undeniably important, but one of the reasons for the species richness of insects is their size. This enables them to exist in small niches, some of which are flowers and other plant reproductive structures. Put simply, they make good pollinators because they literally fit. Other reasons why insects have speciated is the adaptability of their mouthparts and other limbs, and their exoskeleton provides good protection in the exposed environment that is a flower.

LEPIDOPTERA
Most butterflies and moths are pollinators; this hummingbird clearwing (*Hemaris thysbe*) is feeding on lobelia.

Another very important feature that is common to the Big Four insect pollinator groups is that they are holometabolous, which means that they go through a complete metamorphosis, with an adult flying stage and a very different larval stage. One of the perceived advantages of this lifestyle is that it allows a single species to exploit two different habitats at different times of the year. It also means that the adults and the young do not compete for the same resources. As pointed out earlier, the supply of nectar and pollen can be seasonal or even more ephemeral than that, and so an omnipresent pollinator needs a year-round supply. However, if it eats other food when the plant is not in flower, then that solves the problem. (Another way to cope with the ephemeral nature of a food supply is to store it.)

The following sections look in more detail at the various groups of insects that pollinate plants, each examined in decreasing order of the number of species believed to be involved.

ABOVE: HYMENOPTERA

Bees are famous pollinators, especially honeybees (*Apis mellifera*); this one is loaded with pollen.

TOP: COLEOPTERA

Many beetles act as pollinators, but common pollen beetles can cause more harm than good.

Animal pollination of gymnosperms

As seen in Chapter 2, the majority of extant gymnosperms are pollinated by wind, but two groups use animals. These are the early-diverging lineage of cycads, and the more derived odd bunch known as the Gnetales. Study of these two different groups may help us to understand the origins of pollination and also the evolution of the process. When fossils of intact animals preserved in amber are found and there is pollen on them, the value of the evidence and the strength of the ensuing argument both increase.

SAGO CYCAD
Cycas revoluta female cones.

Early evidence

In 2015, researchers reported on fossils found in Myanmar and Spain dating from 100–105 Ma that contained flies with a long proboscis characteristic of nectar-collecting pollinators. The Spanish specimen also included pollen from an extinct group of seed plants, providing excellent evidence of a relationship between a specialist pollinator and a specific plant, and putting a date on that relationship. Furthermore, the plant was not a flowering plant. The extraordinary detail from the fossil shows an animal that could manoeuvre its proboscis accurately. This implies that this animal would be capable of accessing pollen in wide range of plant reproductive structures, and not just those on one species. And if that was not enough, it looks like an animal that would have been capable of hovering, which is a common feeding position today in a range of pollinators, including pollen wasps and hummingbirds. The pollen is typical of that of an insect-pollinated plant, and it resembles pollen from the extinct genus *Exesipollenites*. What this means is that we should not assume that pollinator specialisation is a modern phenomenon; indeed, it may date from well before the extinction of the dinosaurs.

INSECTS IN AMBER
Studying pollinator evolution is difficult, but insects trapped in amber offer unparalleled opportunities. Ancient pollinators preserved this way may still have visible pollen attached.

CYCAD SEED CONES
Cycads like *Macrozamia lucida* open their cones to allow access to insects. This is a male cone, and pollen collected here must be delivered to a female cone of the same species.

Contemporary pollinators

The species that currently pollinate the cycads and Gnetales come from a range of insect groups, including flies, beetles, moths, wasps and, perhaps surprisingly, thrips. This last group is surprising because modern thrips do not look like promising candidates as pollinators, and for many years they were dismissed as simply flower visitors. Thrips feed and breed within flowers and can do a great deal of damage to commercial crops, but 25 per cent of thrip species are now recognised as pollinators. Some of these are very generalist, including the New Zealand flower thrip (*Thrips obscuratus*), which has been recorded on 225 different plant species. It also appears that thrips have been fulfilling this role for more than 100 million years, as more Spanish amber fossils have been found containing female thrips covered in cycad pollen. These ancestral forms of the insect have structures that look analogous to those used for pollen collection by modern bees. Because the relationship between cycads and thrips was clearly already well developed at least 105 Ma, it can be assumed that its origins are much older.

Moths and butterflies (Lepidoptera)

157,338 species, of which 141,604 (90 per cent) visit flowers.

There are many differences between moths and butterflies that are important to pollination, so why are these seemingly disparate groups classed together in one order, the Lepidoptera? One shared feature is the presence of scales and hairs on the wings and other body parts. These scales are partly responsible for the patterning and colouring of the wings, but they are also where pollen gets lodged on the head and front end of the insects when they are supping nectar.

Both moths and butterflies suck nectar through a proboscis, which is a very sophisticated piece of equipment that can be controlled through a combination of muscles and fluid pressure. They also have a small pump in their head to draw the nectar out of the nectary. The length of the tongue generally varies in proportion to both the weight of the animal and its wingspan, and so moths tend to have longer tongues than butterflies (ignoring the countless species of micromoths). The tip of the proboscis is covered with little spines that can be used to scrape plant tissue to release sap and other fluids. This is thought to be one way in which moths get a reward from nectar-less plants.

The interior of the proboscis is very narrow and so the viscosity of the nectar is a significant consideration. If the concentration of sugars in the fluid is too high (more than about 40 per cent sugar), then the poor animals cannot suck it up quickly enough. To solve this problem some species spit on the nectar to dilute it before sucking.

The lepidopterans are not generally strong or long-distance flyers, with 500 m (550 yd) being the limit for some temperate species. Clearly, however, there are some spectacular exceptions to this rule, like the migrating monarch butterfly (*Danaus plexippus*).

Despite their reputation for being average flyers, the moth species that hover rather than land when feeding are, in fact, strong flyers and, in the process, burn a relatively large amount of energy. They have a number of tricks up their wings to help overcome this issue. First, they use their scales and hairs for insulation, which means that they can build up internal body temperatures closer to those of the bird and bat pollinators. Because this temperature increase is temporary, these animals are referred to as heterotherms (as opposed to ectotherms or endotherms). However, they can regulate their temperature. One way they reduce it is by regurgitating nectar onto the outside of their bodies, allowing them to lose heat through evaporation – basically sweating off nectar. They can raise their temperature through rapid muscle movements (a bit like when we shiver), although only animals that weigh more than 30 mg can do this. Being heterothermic and a strong flyer comes at a cost, and means that the animals are more energetically efficient when the air temperature is higher. In the case of the hawkmoths (family Sphingidae), many are nocturnal despite being heterothermic and so rely on a supply of high-calorie nectar.

BUTTERFLIES

Once unrolled, a butterfly's proboscis can reach deep into a flower in search of nectar, which must be of the right concentration for sipping.

MOTHS

Hawkmoths are more often nocturnal than butterflies, and this day-flying individual is an exception. Its long proboscis easily reaches into *Penstemon* flowers.

Moths and butterflies make good pollinators for other reasons. They mostly have very good colour vision at wavelengths of 300–700 nm and in the ultraviolet range, and although some moth species may be unable to see colours at the red end of the spectrum, their colour vison is not as bad as generally assumed. While butterflies have been thoroughly investigated for their ability to discriminate between colours, moths have been left alone because it was assumed that, as nocturnal animals, they do not need to see colours and in any case select white flowers at night. However, this assumption in clearly wrong – after all, we can see colours and yet we also see white flowers better at night. This is another example of the many incorrect assumptions that have been made in the pollination story, a topic revisited in Chapter 6.

THE PROBLEM OF FAST LIVING

Moths that hover when drinking nectar have some of the highest metabolic rates recorded by scientists. The danger associated with this comes from a build-up of reactive oxygen species such as peroxides, which are molecules that cause oxidative harm, including damaging cell membranes. Work carried out on hawkmoths and reported in 2017 shows that for these species the solution to the issue starts with the sugar-rich nectar itself. The hawkmoths use this in a couple of chemical pathways to produce an antioxidant that acts as an electron donor and prevents the damage.

EN ROUTE

Butterflies can remember good food sources and will return to plants visited previously. Some, such as this tiger longwing (*Heliconius hecale*), follow the same foraging route each day.

Moths and butterflies can be very loyal to a particular species – up to 85 per cent of visits by small skippers (*Thymelicus flavus*) are to the plant species they last visited. However, they may be unreliable. Work carried out on an alpine meadow in China revealed that in one eight-year period, the three hawkmoth species that pollinate the orchid *Habenaria glaucifolia* did not show up for five years on the trot, and the thousands of orchids in the meadow just had to sit and wait. Members of another orchid genus, *Gymnadenia*, have a strategy for coping with the no-show of their pollinators. During the day they emit an odour to attract a butterfly, and at night they use a different smell to attract a moth. There are no other changes and the same flower can be pollinated by both, so moths and butterflies clearly have far more in common than the traditional view of their likes and dislikes would suggest. What is not in doubt, however, is that in terms of numbers of species involved, they are at the top of the insect pollinator leader board.

As with their sight, most lepidopterans have good olfactory skills – particularly in the case of moths, which can detect scents over long distances. Nocturnal moths seem to be the best at this and can sniff out a member of the opposite sex several kilometres away. Pheromones are irresistible to some invertebrates and some plants have exploited this shamelessly (for more on this, see page 119). However, scientists have been puzzled for many years how moths discriminate between the odour they want to follow and the other smells around them. Put simply, it seems that they constantly compare the levels of all the odours and then follow the course in which their target scent is increasing in its relative proportion.

Alongside these visual and olfactory skills, butterflies and moths can be quick learners. They can learn to associate good and bad experiences with particular stimuli, and they also have a good memory. For example, they can learn, recognise and remember those plants that are suitable for laying their eggs and for supplying a meal. They can also remember how to find the plant again and can thus navigate via landmarks in the landscape. The New World *Heliconius* butterflies, for example, trap-line the same plants each morning.

BEST OF BOTH WORLDS

The short-spurred fragrant orchid (*Gymnadenia odoratissima*) can modify its fragrance to attract butterflies by day and moths by night.

Beetles (Coleoptera)

386,500 species, of which 77,300 (20 per cent) visit flowers.

As a group, beetles are understudied. This might be because they are more numerous in the southern hemisphere and tropical regions, and biologists have taken a Eurocentric view of the world for too long. Or it might be because they are perceived as ancient and therefore primitive and a bit dull. Although 386,500 beetle species have been named to date, that figure is increasing daily and there may be as many as 3 million once the world beetle catalogue is complete. Whatever the final total, there are undoubtedly a lot of beetles and many of them visit flowers, for better and for worse.

We know that beetles pollinate gymnosperms, both extinct and extant, and we know that many flowering plants are pollinated by beetles. However, it would be premature to assume that the first flowering plants were pollinated by beetles, even if the insects were already performing the job for gymnosperms.

It is fair to say that beetles are currently not the obvious first choice for a pollinator, but they do have their merits. They are well protected by their hard elytra (modified forewings), and they are strong flyers and climbers. They are therefore able to get almost anywhere and will not be deterred by finding another animal at their chosen destination. In addition, they have chewing mouthparts that are efficient but generally lack finesse and are not good for reaching into tight places – unless the whole beetle is tiny. A few have modifications to their mouthparts, resulting in structures such as pollen brushes and pollen spoons, but these are the exception.

Beetles may have a reputation for being a bit clumsy and blundering in everything they do, but that is unfair. They can be loyal – the *Cetonia* beetles that pollinate guelder-roses (*Viburnum opulus*), for example, will fly 18 m (60 ft) to the next plant – which keeps the plant's genes flowing through its population nicely. Some beetles have a very good sense of smell and some can see colour at close range, but there is no beetle equivalent of the studies into the behaviour of moths and butterflies discussed above.

LOYAL BEETLES
Rose chafers (*Cetonia aurata*) carry pollen from guelder-rose (*Viburnum opulus*) over considerable distances.

Bees, ants and wasps (Hymenoptera)

116,861 species, of which 70,117 (60 per cent) visit flowers.

Lumping bees, ants and wasps together may make taxonomic sense, but they are very different in terms of their contribution to pollination globally.

Bees

The close relationship between humans and bees dates back at least 3,000 years, to a time when the ancient Egyptians floated their beehives up and down the Nile to track the spring opening of flowers. As a result, bees have an elevated status in many biological and economic treatments of the subject of pollination. This is partly because they are fascinating and partly because they have been extensively studied, which in turn has made them even more fascinating. What is it about bees that makes them so special as pollinators? There is no one simple answer to this question and what follows is a summary.

Bees are different from almost all other animals because they use both nectar and pollen for themselves and their offspring. Adults sup nectar and eat some pollen, and the larvae eat pollen and nectar (as honey). Generally speaking, to be a bee is to be a flower visitor – they know no other life. The plants bees associate with have been able to benefit from this need by entering into a close mutualism with the insects, which has resulted in many partnerships.

Bees vary in how wide they will spread their services. A few visit only one species but most will visit several. They might confine themselves to one genus or family, or they might be botanically indifferent. Clearly, some bees are very good field botanists. They can travel dozens of metres or even kilometres in one collecting trip, and they can be very disciplined in their collection protocols, trap-lining and returning to the same plant repeatedly. They can learn routes to areas rich in pollen and nectar very quickly, and they can pass this knowledge on to their cohabitees in the hive if they are social (which is the minority of species). They can even mark flowers as having been visited in order to reduce the chance of a second visit before the flower has

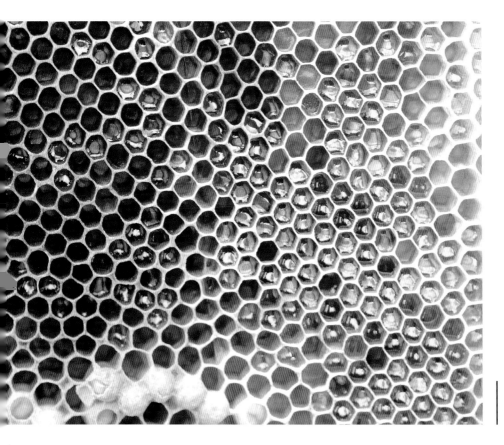

HONEY HIVE
Honeybees gather both pollen and nectar, storing the latter in honeycombs.

secreted more nectar into its nectary. And it appears that they have both short-term and long-term memory (other pollinators may also have this, but bees have been studied intensively).

Bees lap and then pump nectar through a straw-like tube, storing it in a tank in their crop. Also like some lepidopterans, such as the hawkmoths, they can be heterothermic and thus active even on cold days – in fact, a late fall of snow in the UK in April will not deter bees from coming out. Their tongues have a bristly tip for roughing up the food first and separating it, and they may have modified mouthparts that in some species form a scoop.

Bees collect pollen in a number of ways. For a start, there is the simple passive rubbing of the bee's hairy body against the anthers and the subsequent grooming to remove the pollen. The majority of bees are more proactive and deliberately gather pollen. Sometimes this is achieved by scrabbling, but the most spectacular method is buzz collection, where a muscular movement similar to shivering shakes the pollen from the anthers. In this case the bee hangs on very tight to a special cone-shaped structure in the centre of the flower and vibrates it. The plants that are subject to this method have anthers with pores, through which the pollen is then shaken out. This arrangement actually gives both the plant and the pollinator some control over the dispersal of pollen.

Having collected pollen from a plant, the bee may store it in its crop, or it may have pollen baskets on its abdomen or legs. It grooms any pollen from its body and

POLLEN BASKETS

Bees gather pollen on their legs and return it to the hive for later consumption.

puts it in a basket for safekeeping during the journey, and it either glues together pollen stored on the legs with a drop of nectar or compresses it. There are places on the bee's body where it cannot quite reach during grooming, and often these are where the plant tries to stick some pollen in the hope that it will be transferred to the stigma of the next flower.

KEEPING CLEAN

Bees gather and store pollen from their bodies; any grains they can't reach may pollinate flowers visited next.

Bees have good colour vison, similar to that of the butterflies. They are particularly adept at picking out jagged edges to flowers and also patterns visible only in ultraviolet wavelengths. They can learn new shapes of flowers they have not encountered before, which is why they do not ignore non-native flowers such as Himalayan balsam (*Impatiens glandulifera*) in the UK. They cannot see in the dark, and in times of low light levels they can make mistakes. Their olfactory senses are perhaps 100 times more sensitive than a human's and they can learn scents. One experiment claimed to show that bees can discriminate between more than 1,500 different combinations of plant-derived aromas. However, they seem to detect scents best at short range and visual clues at longer distances, which is the reverse of the hymenopterans.

Clearly, as individuals bees are very well suited for the job of pollinator, but it does not stop there because they can be social, at least to some extent, living in groups of up to 180,000 individuals depending on the species. This means that while some bees are active only when one or a few plant species are flowering, some large colonies provide a year-round service. Nectar can be stored in communal structures, thus evening out the potential stochasticity of the supply.

Added to all of these talents is the fact that there are at least 20,000 (and possibly up to 40,000) different bee species, all with different traits, resulting in a terrifically flexible workforce. The scale of their industry is eye-watering – 1 kg (2.2 lb) of honey is thought to be the result of up to 20,000 hours of work, during which up to 22 million visits will have been made to individual flowers in as many as 50,000 foraging trips.

ULTRAVIOLET SIGNALS

Some flowers like this white asphodel (*Asphodelus albus*) have markings to indicate where pollen and nectar are stored. These markings are not visible to the human eye but can be seen in ultraviolet light (right).

Ants

Ants can claim to be just as fascinating as bees, and the global biomass of this group of insects is allegedly at least equal to that of humans and more than any other similar group of organisms. Put simply, there are a lot of ants. So why are they are not big players in pollination? There are many reasons for this.

Ants are small and generally wingless, and so are not good at transporting items over long distances. Although they are small, they can carry seeds and other objects that are huge relative to their body size. The argument that ants are inherently too small to fit inside flowers is a strange one, because surely natural selection would result in the evolution of flowers that fit them a bit better. In fact, many plants treat ants as a threat and have evolved various physical and chemical defences against them.

Ants are fastidiously tidy and clean off any pollen dusted on them. Pollen rarely sticks to them in the first place, however, because unlike moths, butterflies and bees, they have a shiny body surface that is generally non-stick. Ants often produce antibiotics to protect themselves and their nests from bacterial and fungal infections, and these same antibiotics could kill pollen grains. The final reason why ants are unsuitable as pollinators is that they are mainly carnivores. There are exceptions, of course, such as the herbivorous farming ants and the few species that are omnivores. These ants can often be seen collecting nectar from the flowerheads of members of the carrot family (Apiaceae) and spurge family (Euphorbiaceae). In the case of the latter, the ants are probably selfing the plant, but spurges have a very relaxed attitude to the origins of the pollen that is presented to their stigma and welcome anything from the same species – and often even that requirement is waived.

A further reason why ants are not frequently found in the role of pollinator is their avoidance of eating in the open, where they are exposed to predators. The insects are well known as seed dispersers, and many plants attach a fat-filled elaiosome to their seeds to induce ants to take them away from the parent plant. Ants will choose to pick up the whole seed, including the elaiosome, and carry it to their nest rather than spending time in the open removing the elaiosome and taking just that home. It could be that they are happier to act as pollinators for orchids (see box) because the pollen of this group of plants is in sticky lumps the size of seeds, and can be taken quickly in one swift visit to the flower.

BELOW TOP: LOW BLOOMS

As ants rarely fly, they are unsuited to pollinating tall plants, but nailwort (*Paronychia argentea*) blooms at ground level.

BELOW BOTTOM: ANT NEIGHBOURS

The parasitic plant *Cytinus hypocistis* occurs in similar Mediterranean habitats to nailwort and also makes use of ants as pollinators.

ANT POLLINATORS

The fact that ants do not generally fly and are small means that they could still pollinate plants that carry their flowers at ground level. One such prostrate, tiny-flowered plant is nailwort (*Paronychia argentea*) in the carnation family (Caryophyllaceae). Several plants often grow through each other, and they produce inflorescences that initially appear very confusing but are beautiful when viewed under a hand lens. The flowers are self-incompatible and so the random kicking around of pollen grains characteristic of ant pollination will probably be sufficient to meet their needs. Growing near *Paronychia* in the same habitat in the Algarve is the parasitic, bright orange-and-red *Cytinus hypocistis*, which is well recorded as being successfully pollinated by ants. In addition, and perhaps not surprisingly, the plants that do actually pull the wool over the eyes of ants and hoodwink them into carrying pollen are some orchids.

CUCKOO WASP
While adult cuckoo wasps consume nectar, the larvae predate the larvae of bees.

Wasps

The term 'wasps' covers a multitude of separate evolutionary branches and twigs. Like some ants, wasps have been deceived by orchids and there is also an extraordinary relationship between fig wasps and fig 'fruits' (see page 152). However, although wasps will visit flowers to drink nectar, they are not collecting for their offspring as in the case of bees, because young wasps are carnivores. Some wasps have shiny, glabrous heads, while others are hairy; of these, the latter may be effective pollinators and the members of one small group, the pollen wasps (subfamily Masarinae), definitely are. They achieve this status by behaving like solitary bees, collecting nectar and pollen, and then making a mush out of this and using it to fill the cells in their nest. They then lay their eggs into the mush before closing the cell, thereby supplying their offspring with a food source once they hatch.

So some flower-visiting wasps are pollinators, but they do not come close to bees in terms of the numbers of flowers pollinated. However, this does not mean that they are irrelevant or that they can be overlooked.

True flies (Diptera)

155,477 species, of which 55,000 (35 per cent) visit flowers.

The last of the Big Four groups of insects that pollinate flowers are the true flies, encompassing biting midges, fungus gnats, hoverflies, fruit flies, blowflies and house flies. These may not be charismatic on the whole, but they include a significant number of pollinators, not least one species that pollinates the flowers of the cacao tree, the source of cocoa and chocolate. The problem with dipterans is that they are a very diverse group, and so while they share a long list of suitable traits, some are beetle-like and some are wasp-like in their appearance and foraging behaviour. However, one advantage that they can have is being active at times of the year when there are fewer plants in flower. Because they do not feed their young and because they are lightly built, they do not need much provisioning and thus a small pollination reward is generally sufficient. Flies have good visual skills and can be strong and agile flyers.

Other invertebrate pollinators

A number of other invertebrates have been identified as pollinators, including termites. The earliest known proof of termite pollination is in the form of an amber fossil found in the Dominican Republic. Here, a termite was caught in tree resin about 30 Ma while carrying a mass of pollen that is characteristic of the milkweeds (Asclepiadaceae). The best current example of termite pollination is seen in Western Australia, where species in the genus *Drepanotermes* pollinate the western underground orchid (*Rhizanthella gardneri*, see box on page 105).

TERMITES IN AMBER
Although termites are not common pollinators today, fossilized amber shows that the insects have been pollinating plants for millions of years.

Chapter summary

The role of pollinator is one that has been undertaken successfully by a wide range of animals. They are all capable, at the very least, of transporting pollen from the male parts of a cone or flower to the female parts of the same species. More often than not, many animal species pollinate a plant species, and more often than not, many plant species are pollinated by the same animal. For this relationship to work, the same animal has to be persuaded to visit a flower when it is in its male or female phase, and then visit a flower of the same species in the opposite phase. The plant achieves this by first attracting the animal's attention and then rewarding the animal for its visit. Chapter 4 looks at the various ways in which plants attract animals, and Chapter 5 focuses on the many rewards on offer.

Chapter 4

Attraction

As seen in Chapter 3, there are hundreds of thousands of animals with the potential to fulfil the role of pollinator to some of the 300,000-plus plant species in need of accurate, timely and reliable transport for their pollen. It is one thing to have the talent, however, but quite another for the employer to be able to tell the employee where and when to deploy that talent. For this to happen, there must be a form of communication. When it comes to pollination, this communication takes the form of one-way signalling. This chapter looks at the signals put out by plants in order to control their pollinators.

Large tiger hoverfly (*Helophilus trivittatus*).

What is communication?

To behavioural biologists, who are almost invariably zoologists, the words 'communication', 'signal' and 'cue' all have very precise meanings. Unfortunately, however, these meanings can vary depending on which scientific discipline is using them.

Communication occurs between a sender and a receiver. The sender stimulates the sensory systems of the receiver and there is a change in behaviour of the latter. In this definition there does not have to be intent on the part of the sender, which is just as well because plants are not sentient organisms. However, even this is now up for debate because some researchers think there is an action potential across the plumbing of plants that could allow them to assess signals and prioritise their response. Watch the scientific literature for more details of this!

A cue is a form of communication that has no intent and has not been favoured by natural selection for a purpose. An example of this would be the scent a bee sometimes leaves on a flower – this is just some molecules that are rubbed off the outer surface of the insect, much like scent that might be left on a pillow. Animals that come along afterwards and detect the scent can assume that a bee has recently visited the flower. Cues can have a value, or an index, and this is inherent in the cue and cannot be changed – for example, the size of an organism.

FLOWER POWER

Flowers signal to insects, such as this hummingbird hawkmoth (*Macroglossum stellatarum*).

SENDING SIGNALS
The colour of this aster (*Aster* sp.) acts as a signal to attract pollinators including hoverflies.

What is a signal?

Signals are traits (structures or acts) of the sender that are perceived by the receiver, with the result that the behaviour of the receiver changes in such a way that, on average, the signal is favoured by natural selection and becomes fixed. Some biologists have become worried by the fact that the creation of signals comes at a cost, and that the net cost is a handicap to the signaller. This means that dishonest signalling, where the cost is reduced because there is no reward, could become fixed. Obviously, dishonesty does not take over and honesty prevails – and this can be proven mathematically. While the signal does come at a cost, it also comes with a benefit that justifies the cost. Just as a car is costly to run but this cost is outweighed by the benefit of getting you to work, so successful reproduction is the benefit to a plant that makes it worth the cost of employing a pollinator.

Signals produced by plants may be acoustic, olfactory, visual or physical. Acoustic signals involve sound (which may or may not be perceived by humans). Visual signals can be much more variable, and include colours, contrast between light and dark, shape and movement. Olfactory signals are scents, which may be specific to the message or may exploit an existing signal such as a pheromone.

Physical signals involve contact, or even the sensation of heat, leading to a reaction or behaviour that results in the discovery of the reward. It is important to say here that in some cases signals and rewards are very closely entwined in a chicken-and-egg fashion, but that signals are described before rewards here because that is the order they come in the process of pollination. (For more on the evolution of signals and rewards, see Chapter 6.)

All the different types of signals are found many times in pollination communications and mostly they are used honestly. But this is nature, and sometimes deceit and selfishness are not only acceptable, they are successful. Signalling theory requires signals to be accurate (the content of the message) and efficient at getting their message over (how the message is transmitted). However, the success of the signal will also depend on the ability of the receiver to detect it and discriminate it from other similar signals, and then also remember what it means. It is very difficult to know what signal a receiver perceives and we tend to assume that it is similar to what we are picking up. However, what is important here is the reaction of the receiver, rather than worrying about whether, for example, a petal looks black or blue.

A complex system

A good example of the complexity of signalling is that plants offer both pollen and nectar to bees. Researchers have shown that, to a bee, yellow or white petals may indicate the presence of pollen, whereas blue, purple or pink can indicate nectar. This is not a ubiquitous rule, however – tower of jewels (*Echium wildpretii*), which blooms at 3,000 m (10,000 ft) on Tenerife, has red flowers with blue pollen that is harvested by bees, after which the flower turns blue.

This phenomenon of changing colour is widespread across the evolutionary tree but rare when it does occur, and so biologists have tried to find a selective advantage to explain it. It may be a purely chemical artefact of the disintegration of the flower, in which case it has no cost. If there is no cost, then there is no need for an associated benefit, but some biologists always want everything to be an adaptation. Suggested functions of colour change include making the pollinators more accurate and loyal. Outcrossing could be promoted if a change of colour resulted in the direction of pollinators first to the older flowers that are ready to receive pollen. When the change is to green, it may be that the flower is now photosynthesising and providing energy for the developing fruit.

COLOUR CHANGE
After pollination, tower of jewels (*Echium wildpretii*) flowers turn from red to blue.

In summary, there is no one reason for colour changes in flowers and no reason is well proven. It is clear that there are changes in scent as flowers age and there may be reasons for this, and the structure of flowers can also change with age. For example, the young flowers of *Clematis stans* have a longer tube and so only long-tongued bees can access the nectar. As the flowers age, the tube shortens and other species can also get to the nectar. This promotes the possibility that pollen from more than one plant will fertilise the ovules in one flower.

ASIAN CLEMATIS
Clematis stans.

PETUNIA FLOWER
Blue is one of the less common colours of pollen.

This raises another complexity in the signalling. As in the example of the tower of jewels above, sometimes the reward *is* the signal – in this case the blue pollen. Another example is scented nectar, in which case the reward is probably hidden in a nectary and so the signal is just a signpost. This is one place where lies can be peddled. For example, if an orchid flower smells the same as a more abundant nectar-rich neighbour, it does not need to produce nectar itself because by the time the bee finds that there is no nectar, it is too late. The orchid is onto a winner here, because as flowers take some time to refill with nectar after a visit, pollinators are apparently not surprised when they find a flower with a signal but no reward. Here, an empty flower may not be a trick – it may just be refilling – and pollinators have evolved to be tolerant of the odd empty flower and lying orchids persist.

Combined signals

In nature, different types of signal can be used individually or in combination, either in parallel or in series. The combination of the different types of signal can help to maintain the diversity of plant species that we currently enjoy, and we need this diversity if we are to preserve the resilience of our world to change. It has been well known for over a century that honeybees (*Apis mellifera*) can learn to combine both colours and scents into multimodal or complex signals. However, the bees may be faced with many more than two potential signals when approaching a plant. For example, visual variables alone include the size of the inflorescence; the arrangement of the flowers; the size, shape and three-dimensional structure of the flowers; the colour(s) of the flower in terms of their hue, tone, tint and shade; and finally the patterns on the flowers. It is also known that bees can prioritise signals. In 2008 a group of researchers at Cambridge University showed that bees place greater importance on a higher concentration of sucrose in nectar over a warmer flower. Some scientists have suggested that multiple signals help to reduce uncertainty in the mind of a pollinator, allowing it to make more accurate decisions and hence reduce time wasted on the wrong flowers.

Things become even more complicated when the same flower on the same plant sends different signals to different pollinators. For example, *Aphelandra acanthus* flowers are pollinated by bats at night, which are attracted by their scent, and by hummingbirds and diurnal moths during the day, which are attracted by their yellow colour. While bats bring far more pollen than the birds and moths, three out of four of the grains brought by the bats were from the wrong species. This might be an example of a plant hedging its bets in a habitat where pollinators are not reliable.

BIRDS AND BATS
Aphelandra acanthus signals to two different groups of pollinators.

What message is being conveyed?

Signals in nature generally convey one of three messages: 'come and see me', 'don't come and see me' or 'I'm not here' (camouflage). To these might be added 'remember me' and 'off you go'. While a plant wants to attract pollinators most of the time, it also wants them to move off soon afterwards to another conspecific individual. And therein lies a conflict. Some tobacco plants (*Nicotiana* spp.) are pollinated by hawkmoths, which are attracted by a scent that includes benzyl acetone. However, when the moth sups at the nectar it discovers that it is laced with unpalatable nicotine. The current interpretation of this contradiction is that by making the hawkmoth leave quickly, the plant is promoting outbreeding.

To summarise this chapter so far, there is an increasing awareness that pollination is rarely a one-to-one relationship. The 'conversations' between plant and animal are 'heard' by a large audience whose members react in a variety of ways. Therefore, there is a network of relationships with connections that become increasingly complex the more you look.

Pollination ecology involves aspects of sensory ecology, behavioural ecology and evolutionary ecology. When the physicist Erwin Schrödinger (1887–1961) was asked to define life in 1942, he prepared his audience for a complex lecture but reassured them at the start that 'the physicist's most dreaded weapon, mathematical deduction, would hardly be utilised. The reason for this was not that the subject was simple enough to be explained without mathematics, but rather that it was much too involved to be fully accessible to mathematics.' Anyone studying pollination will find themselves agreeing with Schrödinger time and time again. In order to bring some order to this examination of attraction, as opposed to rewards, the remainder of the chapter looks at each of the various forms of signal in turn, even though they do not always exist on their own.

KICK THE HABIT

Jasmine tobacco (*Nicotiana alata*) nectar contains nicotine, so hawkmoths visit for only a short time.

Acoustic signals

A great example of acoustic signalling is an ice-cream van tinkling its tune, indicating that sweet rewards exist in the freezer. Clearly, no flower has been heard making a sound for the same purpose. And while it is conceivable that the sound of pollinators partying in the chamber of an arum flower (family Araceae) could attract the attentions of passing animals, there is no proof of this. However, there is plenty of evidence that the sounds bats make as part of their echolocation system are used by the animals to detect flowers, and in some cases to ensure that they approach the flowers from as close to the front as possible.

Bats pollinate the flowers of hundreds of species, and many of these have concave structures that reflect a signal back to the bat during echolocation. It has been shown with flowers of the cup and saucer vine (*Cobaea scandens*) that the frequency combination of returning soundwaves informs the bat where the flower is and which way it is facing. The flowers of some other species have a vexillum, a modified leaf that acts as a reflecting dish. An example is the dish above the flowers of *Mucuna holtonii* in the bean family (Fabaceae) and a very similar structure above those of *Marcgravia evenia* in the family Marcgraviaceae. The dish is erected above the inflorescence when the flowers are ready, so this is a signal that is correct both in time and space. When the bat lands on the flowers of *Mucuna holtonii*, there is an explosion and pollen is fired at the animal, which in return gains access to a decent volume of nectar. The echolocation system of the bats is sophisticated enough for subsequent visitors to detect that the flower is shot and that there is little nectar left. When researchers removed the vexillum from flowers, the number of visitors fell by nearly three-quarters. In a further twist, species of *Mucuna* pollinated by bats that do not use echolocation do not have a vexillum.

CUP AND SAUCER
Flowers bounce sound back towards bat pollinators.

SONIC BOOM
Curved leaves above *Marcgravia evenia* flowers readily reflect bat sonar.

Visual signals

It seems obvious and therefore safe to assume that any plant species wishing to engage an animal in pollination would make its flowers clearly visible. The flowers of the century plant (*Agave americana*) are held in clusters of hundreds on a 4 m-high (13 ft), 200 mm-thick (8 in) stem. The 300 mm-wide (12 in) blooms of southern magnolia (*Magnolia grandiflora*) have been likened to doves sitting on the tree, and tropical trees such as jacaranda (*Jacaranda mimosifolia*) flower before their leaves reappear, as do many autumn-flowering bulbs. However, there is the matter of scale. A small pollinator will have no problem negotiating the canopy of a tropical tree, and so cauliflory, where flowers develop on short woody shoots on the stem of a tree, is very common in tropical species but is not generally found in temperate regions.

Not every animal-pollinated plant has visible flowers (see box), but most do. There is a plethora of ways in which flowers can differ and thereby send out different signals that are received by different animals, which can then perceive and remember these differences.

TOWER OF FLOWERS
Lofty *Agave* blooms are visible for some distance.

FLOWERS BEFORE FOLIAGE
By producing blooms before leaves appear, Jacaranda trees create a more striking display for pollinators.

UNDERGROUND ORCHID

For a plant, making sure that your flowers are visible seems like a no-brainer. and yet this is nature and so making any assumptions is foolhardy. As with so many botanical rules, an exception can be found in the orchid family (Orchidaceae). The western underground orchid (*Rhizanthella gardneri*) of Western Australia spends its entire life below soil level. It needs no light because it is epiparasitic on a fungus that also lives in association with the broombush (*Melaleuca uncinata*). Each orchid flower is only 4–6 mm (0.16–0.24 in) long, but they grow in clusters of up to 120. The flowers are pollinated by small animals that can access the underground chamber created by the bracts around the inflorescence, including *Drepanotermes* termites, Cecidomyiidae fungal gnats, an Alysiinae wasp and a *Megaselia* fly. The flowering is described as highly erratic, so having four generalist pollinators is a good strategy. The signal put out by the flowers is likely to be scent rather than visual, until the pollinator is right on top.

DOWN UNDER
Underground orchid flowers are usually hidden in the soil, so must attract pollinators with fragrance rather than visual signals.

Flower shape

Broadly speaking, a flower's shape can be categorised in one of two ways: it can either be cut into equal halves in more than one plane and so has many planes of symmetry, like a buttercup (*Ranunculus* spp.); or it can be cut in half in only one plane, like a sweet pea (*Lathyrus odoratus*). The former is described as radially symmetric or actinomorphic, and the latter is bilaterally symmetric or zygomorphic. Of course, this is nature and nothing is as straightforward as that – columbine flowers (*Aquilegia* spp.), for example, are actinomorphic but the individual petals and their spurs are bilaterally symmetrical. On the other hand, some members of the daisy family (Asteraceae) appear at first glance to have actinomorphic flowers, but these are actually clusters of flowers and individually some are zygomorphic (the outer whorl) and some are actinomorphic (the central flowers).

The basic division of flowers into these two groups seems to be important when considering the provision of rewards and the accurate removal and deposition of pollen, but more important for attraction and signalling is the overall shape. Flowers with divided edges, or compound inflorescences like daisies, are more easily seen by bees than flowers that are more circular in outline, like buttercups.

ACTINOMORPHIC VS. ZYGOMORPHIC
Buttercup (*Ranunculus* sp.; left) and sweet pea (*Lathyrus odorata*; right).

Flower size

Flowers vary in size by several orders of magnitude, from the 0.5 mm-long (0.02 in) single stamen of the male flowers of Aleppo spurge (*Euphorbia aleppica*), to the 250 mm-long (10 in) petal tube of golden chalice vine (*Solandra maxima*), which is pollinated by bats. Strangely, it appears that the tiny-flowered Aleppo spurge and the gigantic corpse lily (*Rafflesia arnoldii*), whose flowers measure 1 m (3 ft) across, belong to sister families on the evolutionary tree, and some authors even place them in the same family. While a different range of flower sizes fit different pollinators, there are other considerations. For example, bees may have excellent colour and achromatic vision for accurate foraging and navigation, but the resolution of their eyes is not good and they cannot see small flowers until they are quite close. These insects may see the shape of a 30 mm-wide (1 in) flower at a distance of 300 mm (1 ft) and its colour at 100 mm (4 in), but the smaller the flower, the closer the bee has to be to distinguish these characteristics.

When considering flowers and pollination, it is sometimes more helpful to think about reproductive units rather than individual flowers. Daisies are a good example of a group of flowers that are arranged in a structure that looks like an individual bloom, and this is not uncommon. The flowers are almost working like a

ABOVE: GIANT TRUMPET

Massive *Solandra grandiflora* blooms accommodate pollinating bats.

BELOW: BABY BLOOMS

Male Aleppo spurge (*Euphorbia aleppica*) flowers are tiny and almost unrecognizable.

WORKING TOGETHER
Asian skunk cabbage (*Lysichiton camtschatcensis*) flowers cluster together on a rod-like spadix.

mammals, birds or even invertebrates? One example of the differences between humans and insects is in their eyes. The resolution of the compound eye of an insect is poor. This means that it can distinguish patterns on a flower only when it draws near and can see them clearly only once it has landed. This indicates that long-distance navigation and flower identification by insects are unlikely to be influenced by patterns.

At the beginning of the twentieth century, German biologist Jakob von Uexküll (1864–1944) proposed the idea of *Umwelt*. This refers to the environment within the mind of an animal – the world as they see it. Getting into the mind of a bee is problematic, not least because of its size, but a century after von Uexküll published his most important work, zoologists like Lars Chittka at the University of London are making inroads into this area. These scientists are mapping the areas of the bee's brain that code for colour and smell – sometimes referred to as the perception space.

team, with some division of effort – the outer flowers have more of a role in attraction, while the inner flowers are more concerned with pollen production. Another example of this is the arum family, whose members have very small flowers that lack sepals and petals and are held on a rod-like spadix, which is generally enclosed in a leaf-like spathe. The spathe may be highly coloured and therefore have a role in attraction, and it may also have other uses (see page 124). One member of the family, the titan arum (*Amorphophallus titanum*), has the tallest unbranched inflorescence on record at 3.5 m (11.5 ft). There are scores of flowers in this structure, but most memorable is the revolting stench that emanates from inside the vase-shaped spathe. A similarly disgusting smell is produced by the flowers of *Rafflesia* species, including the corpse lily mentioned above. Current wisdom suggests a very simple reason for this association between floral gigantism and foul smells, which is that if you really want to look and smell like a rotting carcass, it is more convincing if you are the same size as a rotting carcass.

Flower colour

If you ask gardeners to name the first thing about a plant they use for identification, they will more often than not say it is the colour of the flowers. This is despite the fact that flowers are one of the most ephemeral traits. What this shows is how powerful colour can be, or rather how important it is to a primate. But is the same true for other

BEE TALK
By viewing marsh marigolds (*Caltha palustris*) in visible light (above) and ultraviolet light (below), we can see what bees see.

COLOUR PREFERENCE
Bees prefer white *Erysimum scoparium* blooms to purple, as they contain more nectar.

What we do know is that colours can attract animals subjectively, and we know that pollinators can differentiate between plants and select those they wish to provide with their pollination services. For example, work carried out by Jeff Ollerton and colleagues at the University of Northampton has shown that bees are more likely to visit the white flowers of *Erysimum scoparium* than the later purple flowers. The latter contain less nectar and may be for signalling over longer distances. So, colour is just one of the factors that informs decisions made by pollinators. Previous experience and/or innate preferences may also be important, as are the shape and structure of the flower. Whether or not these likes and dislikes follow strict taxonomic and phylogenetic lines is looked at in Chapter 6.

How plants make flower colours
Plants make the colours in their flowers using a paintbox of four basic pigment categories: tetrapyrroles (including chlorophylls), carotenoids, flavonoids (including the anthocyanins) and betalains. These are mixed in the plant in the way that artists mix the paints on their palettes to create different colours. Some colours produced by nature defy description, to the point that they look artificial, one example being the greenish blue of the bat-pollinated flowers of the jade vine (*Strongylodon macrobotrys*). This colour (which is very similar to the blue of Cambridge University) is created by mixing two pigments (malvin and saponarin) in a 1:9 ratio at a high alkaline pH, and then combining this mixture with the same two pigments in a different part of the petal at a lower but still alkaline pH. Another example of combining pigments is seen in the potato family (Solanaceae), where purple or blue anthocyanins and orange carotenoids are mixed to make many different reds.

The pigments in flowers are physically located either in cell membranes, in the cell vacuole or in membrane around plastids. There are various chemical ways to make the colour lighter or darker, or more saturated. While most of the colour that we see in flowers is the direct result of light interacting with pigments and being reflected, there are occasions when physics rather than chemistry lends a hand, such as changing the shape of cells on the surface of a petal. Conical cells reflect light differently from flat cells and are reputed to make petals sparkle, which bees apparently like!

Another way to make blue in particular is to use diffraction gratings. These are found on the surface of many petals and produce light predominantly at shorter wavelengths (ultraviolet and blue). The grating consists of nanostructures in almost parallel striations.

COLOUR CHEMICALS
Molecular structure of several compounds creating colour in plants, including the flavone backbone (2-phenyl-1,4-benzopyrone, above left) and chlorin section of chlorophyll a (above middle). The green box shows a group that varies between chlorophyll types. Also shown are betanin (above right) and a carotenoid: polyene tail with double bonds (right).

The 'almost' is important here, as while the striations can be laid down perfectly parallel on synthetic materials, this does not happen in plants. These imperfections enable the plants to produce a range of different blues and ultraviolets that are particularly visible to bees. This means that although a flower may have a base colour that we see, the bees see the blue halo, and it explains why bees pollinate flowers of colours they should not be able to perceive clearly. This research, based at Cambridge University and reported in *Nature* in 2017, involved a wide range of scientists, including zoologists specialising in bee behaviour, botanists specialising in flower structure and material scientists.

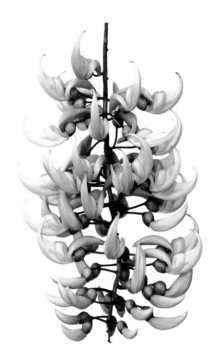

MIX AND MATCH
Combining pigments allows the jade vine (*Strongylodon macrobotrys*) to create this unworldly colour.

THE PROBLEM WITH BEING GREEN

Before leaving colour, green needs to be mentioned. The majority of nature is green – it is the canvas on which life is painted. Indeed, green is so ubiquitous that many gardeners do not even consider it to be a colour. Against this verdant backdrop, flowers are generally visible even if the plant is not in leaf. It is believed that some colours such as yellow stand out particularly strongly against green to vertebrates with colour vision, but not in the case of animals with compound eyes. Likewise, humans and other primates perceive red and green as contrasting colours because of the way in which the information from our eyes is processed in the brain.

It is obvious that plants that produce green flowers are not relying on the contrast between flower and leaf to attract a pollinator's attention. The shape of the flower might do the trick, or if the petals are green then another part of the flower could stand out – the anthers of the kangaroo paw (*Anigozanthos flavidus*), for example, are bright orange against green petals. However, scent is a more likely attractant in such cases. One example of a plant that uses this strategy is *Deherania smaragdina*, which produces flowers that perfectly match the colour of its leaves. Unfortunately for humans, the flowers smell like the inside of a teenager's trainers!

GOING GREEN
Green flowers rely on scent or contrast to attract pollinators; in kangaroo paw, orange anthers stand out.

Coloured flowers

The majority of the colour we see in flowers is in the petals, and yet all the other parts can be coloured (including in the ultraviolet spectrum) and contribute to the signals or cues.

SEPAL SWAP
Hellebores utilise colourful sepals rather than petals.

PRETTY PERIANTH
Water lilies have vivid petals and sepals.

Calyx

Working from the outside in, the outer whorl of a flower is the calyx, which often comprises green or brown sepals and has a protective function. However, the sepals of hellebores (*Helleborus* spp.) are described as petaloid, which means they are coloured and conform more to our idea of what a petal looks like rather than a sepal.

Perianth

The sepals and the petals together make up the perianth, although sometimes this is not differentiated into sepals and petals and both are coloured. Many varied species have this arrangement, such as water lilies (family Nymphaeaceae), magnolias (*Magnolia* spp.) and tulips (*Tulipa* spp.). Some flowers do not have a perianth, and if they are pollinated by animals they therefore have to find other ways of signalling.

Bracts and leaves

Some plants have leaves or bracts that contribute to the colour, such as the upper stem leaves of poinsettia (*Euphorbia pulcherrima*), which are red. Surrounding the tiny flowers of the explosive creeping dogwood (*Cornus canadensis*; see page 28) are four white bracts that are often mistaken for a perianth, and it has to be acknowledged that they are fulfilling the role usually played by petals. Above the very strange pendulous flowers of the handkerchief tree (*Davidia involucrata*) are two uneven white bracts. It would be sensible to assume that these are also for attracting the pollinator, and when they are removed there are far fewer visits from the pollinating bees. However, the bees also see the anthers very clearly, and the flowers smell like a territorial tomcat, so botanists surmised that these bracts might have another role. Research revealed that the tree's pollen is easily washed away by rain and will germinate prematurely if it gets too wet, and that the bracts act as an umbrella, keeping the pollen dry.

Stamen

Other flowers that lack a perianth have different parts that are coloured. The Persian ironwood (*Parrotia persica*) has a superfluity of bright red stamens, while its close relative Chinese fighazel (*Sycopsis sinensis*) has orange-yellow stamens and a much-reduced brown perianth.

BOYS IN BLOOM
Persian ironwood flowers lack petals; their male stamens create a show.

Pollen

The pollen itself can be very colourful. Mention has already been made of the orange pollen of lilies (see page 26) and the blue pollen of the tower of jewels (see page 100), but there are many other shades and hues. The majority of pollens appear either yellow or white to the human eye, but to a bee they look quite different – it has even been shown that bees can discriminate between species based on pollen colour alone. Researchers have also observed that bees seem to show a preference for visiting flowers where the pollen is the same colour as the centre of the flower.

Pollen can give chemical cues as well as a colourful signal, and bees can learn to associate these with high- or low-quality rewards and amend their behaviour accordingly. Bees have also been shown to check constantly that they are making good choices. Innate preferences seem to be quickly overwritten, but different bees can show preferences for different colours of pollen, and so pollen of several colours can be maintained in the same species of plant.

Stigma and style

It is rare for the female parts in the centre of the flower to be colourful, but in some – including those of wild angelica (*Angelica sylvestris*) and other members of the carrot family (Apiaceae) – they can appear very bright when viewed under ultraviolet light.

Nectary

In reality any part of a plant could potentially be selected to attract a pollinator visually, including the nectaries. The inflorescences and flowers of spurges (*Euphorbia* spp.) are strange and have confused many people, including the Swedish taxonomist Carl Linnaeus (1707–1778). The flowers are small and unisexual, but they are spatially arranged in a cup like the male and female components of a more typical flower such as a geranium (*Geranium* spp.). Around the rim of the cup are nectaries. These can vary greatly from species to species, and in some there are petal-like outgrowths from the edge of these glands, making the inflorescence look even more like a 'normal' flower.

> **NECATARIES ON SHOW**
> Row 1 left to right: *Euphorbia quadrilatera, E. guiengola.*
> Row 2 left to right: *E. baylissii, E. neococcinea.*
> Row 3 left to right: *E. mafingensis, E. awashensis.*

Nectar

Nectar can range in colour from clear like water to yellow, amber, pink, red, black and blue, with the most common colours being yellow through orange to red. In some of the 67 taxa in which coloured nectar is recorded, it contrasts with the rest of the flower. The distribution of coloured nectar on the evolutionary tree appears to be quite random, but it tends to be associated with three very different situations: bat and bird pollination on islands or otherwise isolated regions, and in locations at high altitudes. One intriguing example of coloured nectar is from the island of Mauritius, where the endemic *Phelsuma* geckos prefer blood-red-coloured nectar. Coloured nectar is an honest cue for the pollinator and therefore reliable; it is not mimicking a deterrent. So the presence of the coffee-coloured nectar in a giant honey flower (*Melianthus major*) is a safe bet for the gecko. It has also been proposed that the coloured nectar is a deterrent to nectar thieves; some pollinators simply do not like the look of it.

Pollination drop

The pollination drops of gymnosperms have been compared to nectar in species where animals pollinate the reproductive structures. Perhaps the strangest and most romantic example of this is the enigmatic genus *Ephedra*. These plants look like a shrubby form of *Equisetum* (horsetails), and although they produce reproductive structures that could be mistaken for flowers, they sit fair and square in the conifers. In 2015, a Swedish research group reported the unexpected correlation between *Ephedra* pollination and the phases of the moon, and in particular the July full moon. They suggested that part of the reason for this correlation is that the pollination drop is more visible to the nocturnal dipteran and lepidopteran pollinators, and that the fluid glistens in the moonlight, thus catching the eye of the pollinators.

DANCING IN THE MOONLIGHT
The pollination drop of the gymnosperm *Ephedra funerea* glistens under moonlight, improving pollination during the full moon.

Olfactory signals

The odours that plants produce in the name of pollination vary from the sublime and wretch-inducing to undetectable to humans. When you see an old-fashioned rose like *Rosa* 'Zéphirine Drouhin' or a Madonna lily (*Lilium candidum*), you have a very strong urge to smell the flower if you have smelt one before. Likewise, if you take a walk through a botanical garden in February you are very likely to get a whiff of the extraordinarily strong scent of sweet box (*Sarcococca* spp.), and even if you have never seen nor smelt the plant before you will seek it out. Scent is powerful – childhood memories can be evoked almost instantly by a particular scent. Smells are judged in context by many animals, not just humans, meaning that the same molecule can mean different things in different locations. Plants exploit the ability of animals to differentiate between smells and to remember them. Not only that; the animals can remember what the smells remind them of, which may be nectar, pollen or any of the other rewards offered by plants (see Chapter 5).

HEADY SCENT
Madonna lily (*Lilium candidum*) flowers have a potent fragrance.

Chemical cocktails

So, smells (pleasant to humans) and odours (unpleasant to humans) are a signal and/or a cue in much the same way as colour, but there is a problem in the study of odours that has undoubtedly held up their investigation. While artists and garden designers bemoan the fact that colours are difficult to describe accurately, they are a doddle compared to describing smells. Ironically, in a seminal work published in 2008, American biologist Robert Raguso had to resort to using colour-coding for the different classes of molecules to make his excellent diagrams clearer, and this approach has been adopted in many subsequent papers – after all, words like 'flowery' and 'spicy' are not very helpful. The leaves of the giant honey flower have been described by different people as smelling of either vomit or peanut butter – a case of beauty being in the nose and the eye of the beholder.

FAR-FLUNG FRAGRANCE
The scent of winter-flowering honeysuckle (*Lonicera fragrantissima*) travels widely.

While colour has been considered in detail for centuries, the study of scent is way behind, though there has been a resurrection of interest in the subject at the start of the 21st century with the work of Raguso, along with Steven Johnson from South Africa, Florian Schiestl from Switzerland and others. Why scent should be so much more difficult than colour to investigate and discuss is the result of many different factors. Undoubtedly, identifying the molecules can be very expensive.

An even more important reason why investigating pollination scents is so difficult is that scents are, more often than not, cocktails of several to many molecules. Where one molecule only is used, the plant–pollinator relationship tends to be more specific. For example, sulphur compounds such as dimethyl disulphide are almost always associated with attracting bats or animals that have an innate liking for carrion. However, most scented flowers use a mixture of between 20 and 60 components, and up to 150 different scents have been isolated from sunflowers (*Helianthus* spp.). While perhaps only 30 of these are recognised by pollinating bees, the complexity of the floral bouquet is staggering. This is further magnified by the opportunity plants have to mix the same cocktail of chemicals in different ratios. Pigments are never combined in similarly complex mixtures.

SWEET SCENT
Honey flower (*Melianthus major*) flowers live up to their sweet name.

CLASSIFYING FLOWER SMELLS

The smells flowers produce derive from relatively small amounts of volatile organic compounds (VOCs). These are often small carbon-based molecules that are relatively cheap for plants to produce, and that can cross cell membranes easily and then evaporate, thus spreading the message. There are about 1,700 named molecules involved in flower smells (and no doubt many others yet to be named), of which the principal types are aliphatic compounds (hydrocarbons, alcohols, esters and ketones), terpenoids (monoterpenes and sesquiterpenes) and benzenoids. An alternative classification of the chemicals in the flower scents is the more subjective approach: 'flowery' scents include terpenes, benzenoids and some alcohols, ketones and esters; dung/carrion smells include amines, ammonia, indoles and the iconic skatole; the pheromone-mimicking molecules includes alcohols, acids, esters and terpenoids; and some of the relatively uncommon but well-documented scents of bat-pollinated flowers include sulphur-containing compounds. The point here is that these words are not as familiar to most people as the names we have for colours – after all, we do not talk about colour by naming the pigments.

STRUCTURES OF SCENT
Four common fragrance compounds are ethanol, acetone, zingiberene and coronene (clockwise from top left).

The many roles of pollination scents

Another difference between visual colour signals and scent is that the molecules that produce scents have more than one role to play. These roles can be divided into those directly related to pollination and those with an indirect effect. The latter may be a hangover from a former role and so might be useful when we are trying to understand how pollination evolved.

Taking the direct roles of scent in a chronological order in relation to pollination, they can first be used for long-distance attraction, as in the case of moths. In this particular instance the scent is produced at the time of day the moths are active. This switching on and off of scent production helps the plant to save energy and make the signal a bit more specific. Next, scents may be used for short-distance attraction and orientation, such as indicating to bees that it is time to land. Having landed, the bee might be prompted to forage deeper into the flower or to extend its proboscis by following the distribution of different scents as a directional nectar guide (for more on this, see the box on page 121.) Finally, and perhaps counterintuitively, the scent can be used as a sign to the pollinator that it is time to go away and find another plant, thus promoting outcrossing (see box).

LONG-DISTANCE COMMUNICATION

Fragrance attracts pollinators from further afield; this hummingbird hawkmoth (*Macroglossum stellatarum*) has found purpletop vervain (*Verbena bonariensis*).

CYCAD SCENTS

A fine example of a plant releasing different scents over time comes from Australia, where the cycad *Macrozamia lucida* is pollinated by *Cycadothrips* thrips, which eat the pollen in the male cones. The cycad sets off the initial attraction by producing relatively low levels of β-myrcene and (E)-β-ocimene, but around midday the cones warm up and there is a million-fold increase in the levels of these scent chemicals. The thrips cannot tolerate these raised concentrations and so move on to the female cones (on another plant), which emit relatively low levels of β-myrcene and (E)-β-ocimene. Note that having said that specialised relationships tend to involve specialised molecules, β-myrcene and (E)-β-ocimene are two of the most commonly used scents. Never expect nature to stick to any rules!

The indirect roles and side effects of scents are very varied. There is good evidence that caffeine in nectar can improve the memory of honeybees, which then have a greater chance of finding their way back to the plant and its conspecifics. This is a dangerous trick for the plant to play, however, because too much caffeine will act as a deterrent. Some of the other uses to which the scent molecules are put include defending against microbes, facilitating heat loss and reducing damage from oxidative stress. Unfortunately, some non-pollinating flower visitors, such as crab spiders (family Thomisidae) are attracted by the same scents that attract pollinators. The spiders sit and wait for the pollinators to turn up and eat them when they do. The scents produced by *Acacia* trees appear to repel ants when the pollinators are about to arrive.

It is well known that scent becomes stronger and is detectable from a greater distance if it is warmed – essentially, more of the volatile compounds evaporate. This is exploited by humans when we spray perfume on our pulse points – parts of the body where the warm blood vessels run close to the skin's surface. A taxonomically diverse group of plants deliberately heat their inflorescences, achieving this simply by orienting their flowers towards the sun and even tracking it, or by burning lipids to generate heat in what is basically unregulated respiration. The same heat that is generated by these plants may also be part of the reward for the pollinator (see Chapter 5).

POLLINATOR IS PREY

Flower fragrance attracts bees and other pollinators, but also predatory crab spiders. Their colourful bodies blend well with the blossom, providing perfect camouflage as they hunt for insects; here, a honeybee (*Apis mellifera*) has succumbed.

HOT NECTARIES

The Christmas rose or black hellebore (*Helleborus niger*) commonly flowers while snow is still on the ground. The heat generated by the nectaries can help to melt the snow around the plant.

Where is the smell coming from?

While it is tempting to assume that the scent coming from a plant is produced by its flowers, this is not always true. Rosemary (*Rosmarinus officinalis*) has highly scented leaves and very pale blue flowers that are very difficult to see on a sunny day. Yet even on Cape St Vincent in Portugal, one of the windiest places on Earth, bees find the flowers. It is widely believed that scent from the leaves has many roles, including attracting pollinators. In the same habitat is French lavender (*Lavandula stoechas*), which has three large, light purple bracts at the top of an inflorescence that consists of many tiny, dark purple flowers. While it is tempting to assume that the bracts are there to help attract pollinators, removing them appears to have no effect on seed-set. Again, the scent from the leaves seems to be the major attractant. A third plant from the Algarve that uses leaf-derived scent is the dwarf palm (*Chamaerops humilis*), in which the widespread chemical (E)-β-ocimene is released from the leaves around the inflorescence. While this might attract a range of mostly inappropriate pollinators, only the weevil *Derelomus chamaeropsis* (which breeds in the plant's male flowers) can gain access through the maze created by the smelly young leaves and pollinate the flowers.

NON-FLOWER FRAGRANCE

Kitchen herb rosemary (*Rosmarinus officinalis*) has fragrant leaves but scentless flowers (top).

Dwarf palm (*Chamaerops humilis*) leaves release a scent that attracts weevils towards their yellow flowers (above).

Derelomus weevils pollinate dwarf palm while burrowing into the inflorescence stalk to lay their eggs (left).

It is now clear that the different parts of a flower smell differently, although it is worth noting that this assumption is based on a relatively small amount of data. While nectar can be coloured (see above), it can also be scented. This scent is an honest signal, because it is being released from the reward that the pollinator is seeking. The nectar may acquire its scent in one of three ways: the scent molecules might simply be diffusing out of the petals; the scents might be synthesised specifically for inclusion in the nectar (this must be the case if the molecules are found only in the nectar); or the scent may be produced by yeast and fermentation later on in the process of flowering.

How is scent selected?

There are two routes by which a relationship can be established between a scented flower and a pollinator. Research has shown that bees are capable of learning to associate new scents with existing rewards, thereby creating new relationships. However, pollinators may already have had an existing relationship with the scent before the plant barged in. For example, the beetles that pollinate members of the arum family are attracted by VOCs emitted by the inflorescence. However, those beetles were already using the VOCs for communication with each other for millions of years before the plants started synthesising a reward molecule with the same VOC scent.

A similar situation is seen when the plant puts out a widely existing scent that pollinators already associate with a reward. The classic example of this is the scents that mimic dung and faeces. Flies find these irresistible, and many different groups of plant have evolved the ability to produce skatole and similarly unpleasant-smelling compounds. Whether the plant provides a reward in these cases is investigated in Chapter 5.

Generally speaking, a plant has very little control over which organisms perceive the signals it gives out from its leaves, flowers, cones or fruits. Gardeners can admire the beauty of flowers – their shape, colours and scents – but the plant does not benefit in any way from this (except that the gardener is more likely to cultivate the plant). However, sometimes a plant starts to produce a scent that is meaningful to only one species. These relatively rare but well-documented monolectic relationships are generally based on very specific pheromone mimicry, with the plants exploiting a pre-existing association in a deceitful way. The poster plant for this selfish behaviour is the Chinese orchid *Dendrobium sinense*, which produces the same scent as that given off by a bee in the genus *Apis* when it is being attacked. This does not attract conspecific bees to come to the assistance of the fake bee, but the hornets that prey on the bees. When the hornet lands on the orchid looking around for dinner, it acquires pollinia (masses of pollen grains stuck together) from the orchid. This is a very specific line of communication, referred to as a private channel, and is based on the idea that the bees' alarm scent means nothing to any species other than their conspecifics and, unfortunately, their predators.

UNDER ATTACK
The orchid *Dendrobium sinense* tricks hornets into visiting its flowers.

On the scent

Among the few detailed studies of scent emissions is one reported in 2010 by Sandy-Lynn Steenhuisen and colleagues at the University of KwaZulu-Natal, who looked at four *Protea* species that are pollinated by beetles.

They identified 118 different compounds in the scent of the flowers, but the identity and proportions of these varied greatly across the four species. One of the species, the cluster-head sugarbush (*Protea welwitschii*), was noticeably different, with several unique scent compounds, and its nectar contained 55 different compounds. The proportions of the different compounds changed during the lifetime of the inflorescences, from opening through the release of pollen (anthesis) and as the inflorescence went over. Monoterpenes and aromatic ethers made up the majority of the emissions, and the greatest similarity between the species was at anthesis, when linalool (described as floral with a touch of spice) and benzaldehyde (almond-like) made up most of the scent.

Different parts of the inflorescences (bracts, perianth and styles) emitted different combinations of different scents at different times, and the combinations varied between the species. However, the peak of nectar scent emission in all four species was at pollen release, and the scent of the nectar was similar across all four species, supporting the idea that it is produced for the manipulation of the beetle pollinators. The research group believed that the temporal and spatial variation of scent production indicates that scents have several roles, including long-distance attraction of the beetles at the right time. Once the beetles have arrived, then the different scents provoke the insects to rummage around in the inflorescence, thereby pollinating the flowers and getting covered in pollen to take to the next plant.

DIGGING IN

Cluster-head sugarbush (*Protea welwitschii*) flowers are densely clustered, so successful pollinators must root around.

Physical signals

The flowers of some plants direct their pollinators by physical and tactile means rather than visual or olfactory signals. While these are perhaps not strictly attractants, they certainly help the pollinator to find the reward accurately, with the potential benefit that the plant will be pollinated by an animal that is loyal to the species. One example of this phenomenon is seen in the flowers of *Datura* species and their pollinator, the tobacco hornworm moth (*Manduca sexta*), whose proboscis fits into a groove on the surface of the petals. The moth is initially attracted by scent and colour signals, but the final part of the forage is directed by the grooves.

GUIDANCE REQUIRED
Jimsonweed (*Datura stramonium*) pollinators are large-bodied moths, in this case the pink-spotted sphinx moth (*Agrius cingulata*). They do not begin visiting the flowers until the sky is truly dark, more than half an hour after sunset. The moths are attracted by a strong, fragrant aroma. Note the cream-coloured pollen on the lower two-thirds of this moth's proboscis.

NECTAR GUIDES

The *Datura* adaptation described above is an example of a nectar guide, but these are not just mechanosensory in their action. A well-documented example of visual nectar guides is the coloured lines on the petals of some flowers that are visible only to animals that can perceive within the ultraviolet part of the spectrum, such as bees. Nectar guides can also be olfactory, which may help explain the spatial variation of scents within flowers. The variation creates a heterogenous sensory landscape, guiding the pollinator towards the nectar. This has been clearly demonstrated in daffodil flowers (*Narcissus* spp.), in which the corona is formed by the fusion of the stamen filaments. This structure emits the majority of the scent in the flower and thereby brings pollinators closer to the anthers and the stigma.

FOLLOW THE FRAGRANCE
The trumpets of daffodils produce their own scent, thus guiding pollinators towards the crucial sexual structures within.

The whole package

In 2004, Robert Raguso wrote a short but very important and highly cited paper that coined the expression 'flowers as sensory billboards'. The simple message here is that many plants do not use visual and olfactory signals in isolation. Two years later, in a review of floral pigments, botanist Erich Grotewold stated that 'our understanding of the specific cues perceived by pollinators in a total floral landscape is very rudimentary, and other factors, in addition to flower pigmentation, [are] likely to play a significant role'. This is very true. Many plants do attract their pollinators by creating a sensory landscape in which many signals combine to control the pollinators.

This multimodal communication is not surprising when you think that the signals are being used for several different purposes: some are concerned with navigation, some are used for identification, some are attractants and others are used for rejection. The way in which they are perceived is also very different. Visual signals are perceived by just a few different receptors, whereas there are dozens of receptors for smells. A further layer of complexity is that the landscape changes over time – on a shorter timescale than the four seasons, but with a definite purpose nonetheless.

One of the most beguiling pollinations involves the deception of thynnid wasps by wasp orchids (*Chiloglottis* spp.), hammer orchids (*Drakea* spp.) and elbow orchids (*Arthrochilus* spp.). The wasps display an extreme sexual dimorphism, with the females being flightless. The young female wasp will climb about 25 cm (10 in) up a grass stalk and then gives off a pheromone, sometimes while rubbing her legs together. The male wasp flies to the female, plucks her from her grassy perch and, in mid-air, mates with her. The orchids have all evolved a modified labellum petal that looks close enough to the wingless female wasp, and they fool the male wasp into trying to pick 'her' up. In order to make the illusion more convincing, the orchid labellum gives off a molecule that mimics the pheromone produced by the real female wasp.

REDUNDANT FEATURES

For hundreds of years biologists have attempted to describe and explain the relationships between plants and their pollinators. Sometimes it is possible to be drawn into believing that everything about a flower or inflorescence has some role to play in the attraction or reward of a pollinator. However, this is not always the case. Researchers have shown, for example, that the bracts atop a French lavender inflorescence are currently unimportant in pollinator attraction. Another example of a worthless or perhaps redundant feature is the black central floret in the white inflorescence of the wild carrot (*Daucus carota*). This flower is so conspicuous and ubiquitous that it is tempting to think that it is a dummy flower to encourage flies to visit the plant, but there is no experimental evidence to support this.

ROYAL BLOOD

Wild carrot (*Daucus carota*) is also known as Queen Anne's lace. The flowers resemble lace and the single reddish bloom is said to be a drop of Her Majesty's blood from a needle-pricked finger.

WARTY HAMMER ORCHID
Drakaea livida.

A different pollination relationship involving sight and smell is between Bentham's cornel (*Cornus capitata*), a tree that grows in the woodlands on the lower slopes of the Himalayas in China and India, and bees (particularly those in the genus *Anthophora*), along with a few dipterans and lepidopterans. The individual flowers of the tree are small and borne on a spherical receptacle, with the entire ball of flowers subtended by four white bracts. The latter are commonly mistaken for petals, which is not unreasonable because they contrast strongly with the green leaves and could therefore attract the attention of a pollinator. In 2014, a research group from the Kunming Institute of Botany set out to find whether the bracts are important for attracting a pollinator, whether the flowers emit any scents (other species of *Cornus* do this), and whether the visual and olfactory signals are of equal importance.

The researchers detected 12 different scents in the air around the flowers (including linalool and E-β-ocimene), and these were used in a variety of tests designed to answer the other two questions. They found that visual signals were more effective for attracting the pollinators than scent, but that scent was more effective for provoking flower landings. However, they also found that combined visual and olfactory signals were more effective in provoking approaches and landings. It is therefore assumed that visual and olfactory signals work synergistically in intact inflorescences in this species.

The male wasp assumes that this thing resembling a female, at the right height from the ground and smelling like a female wasp, actually *is* a female wasp. It lands on the labellum and tries to take it away. At this point the hinge at the base of the labellum comes into play, and the labellum plus amorous wasp swing forward, bringing the wasp into contact with the sticky pads at the base of the stalks that are attached to the orchid pollinia. The male wasp then gives up and flies away with pollinia attached. As he does so, the stalk of the pollinia bends through 90 degrees, moving the pollinia to a slightly different place on his body. Hopefully, for the sake of the orchid, the male will fall for the deception again, only this time the pollinia hit the stigmatic surface of the second orchid as he swings forward. The response of the pheromone is so ingrained that the male wasp will even take away a metal disc if it is the correct size and shape, it is suspended at the correct height above the ground and it smells of female wasp.

BOUNTIFUL BRACTS
Bentham's cornel (*Cornus capitata*) uses petal-like bracts to attract pollinating insects.

FLY TRAP
The blowflies that pollinate dead-horse arum (*Helicodiceros muscivorus*) receive no reward and some may even die while trapped within the floral chamber.

being used to amplify the olfactory signal and perhaps attract flies into the chamber, but in some cases it can also be part of the reward (see Chapter 5).

Bat-pollinated flowers are some of the easiest to understand because they have a tendency to share suites of characters that enable bats to find them and they have a very big reward for the successful animal. Flower shape has been shown to be instrumental in helping the pollinators align with the front of the bloom using echolocation (see page 103). However, the environment can become so cluttered that the bats get mixed messages from their echolocation, and in these situations scent is more reliable and used preferentially by the animals. Here, a combination of sound, scents and/or visual clues is therefore being used.

One final example of a plant that uses a combination of signals is the Rangoon creeper (*Quisqualis indica*). This produces flowers that are white at first, in the night, but then fade quickly through pink to red the following day. Concurrent with the colour changes is a change in the scent. The list of scent compounds does not change, but their proportions and contributions to the final bouquet do. Nectar is produced through the flowering but the amount produced falls off as the white flower turns red. With the change in colour and scent is a change in the guild of pollinators, from moths to butterflies and bees. This species is self-incompatible and it may have adopted its complicated signalling strategy as a way to exploit a range of pollinators and so reduce the risk of being ignored.

Another example of olfactory and visual signals working together is the dead-horse arum (*Helicodiceros muscivorus*), which both looks and smells like the back end of a horse, complete with hairy tail. Blowflies (family Calliphoridae) fall for this illusion and enter the spathe, out of which the fake tail grows. In this instance, the shape, structure, colour and smell all work together to draw the flies into the spathe, where the very small male and female flowers are housed. The scent is rich in dimethyl disulphide, which is regarded by some as the universal carrion odour.

The arum's female flowers are receptive before the male flowers produce their pollen, but the flies are retained long enough to be present when the pollen is finally released, at which point they are set free. When I have grown this plant in the UK, the chamber not only contains dead flies but also hungry maggots looking for food. Their parents have been deceived into mistaking the flower for a corpse that could support their offspring, but are unable to escape because of the ring of hairs blocking their way. The release of the scent is timed to coincide with the female flowers being receptive, and the chamber is heated as the odour is being produced to make sure the scent escapes far enough. Here, heat is

RANGOON CREEPER
Quisqualis indica.

Chapter summary

Plants use a wide range of signals and cues to attract and orientate pollinators. The visual and olfactory senses of the pollinators are exploited for navigation to the plants and for their identification. The signals may be sent and received over both long and short distances depending on the environment, the pollinator and the plant. It is very common for more than one signal to be used.

Olfactory signals tend to be more complex than visual signals, and they are less well understood. Olfactory and visual clues often show spatial and/or temporal variation throughout the life of a flower or inflorescence.

The signals are mostly honest, but they can be deceitful. The most honest signals are those where the reward is the source of the signal, such as scented nectar. Rewards are as diverse as the signals the plants use to attract a pollinator, and it is these that are considered in the next chapter.

Chapter 5

Rewards

Superficially, pollination is a form of mutualism, or a relationship from which two organisms benefit – in this case the plant and the pollinator. Yet this is nature, so the only surety is that things will not be as simple and clear-cut as this. Indeed, in some pollination relationships the animal is duped and gets no benefit, and in others the herbivorous insect takes some food but does not provide a pollination service in return. Somewhere in the middle are plant and animal species whose relationship is so tightly entwined that the demise of one party would lead to the extirpation of the other.

Ward's stanhopea (*Stanhopea wardii*) with visiting bees.

Benefits and costs

When it comes to pollination, plants have a job that needs to be done – their pollen has to be delivered to the correct stigmatic surface. Animals have needs too – food and drink, finding a mate and successfully reproducing, and finding a safe, sheltered place to sleep. The job description for the post of pollinator given at the start of Chapter 3 lists a number of different payment options covering the basic needs of any animal. Referring to these payments as rewards implies that the pollinator 'knows' what is going on, which is probably not the case. Despite this, 'reward' is the term commonly used in the literature – another example of an unfortunate word choice hindering the investigation of pollination (See Chapter 6).

It is true to say that many animals help themselves to payments without delivering any pollination services, so 'payments' is also the wrong word to use here, partly because it anthropomorphises the relationships. In nature, the word 'benefits' is more appropriate because there is no guarantee that the animal will fulfil its part of the bargain. The benefits on offer become rewards only once the animal has fulfilled its side of the relationship, generally by accident.

When biologists talk about benefits, they often go on to talk about associated costs. In pollination, the costs to the plant are those that any organism has to accept if it wishes to reproduce sexually – you do not get anything for nothing in nature. Different plants allocate and

HEART-SHAPED TONGUE ORCHID
Serapias cordigera.

HAZEL CATKINS
Pollen is expensive to produce, yet wind-pollinated common hazel (*Corylus avellana*) releases large quantities.

SINGLE FLOWERS
Each tulip (*Tulipa* sp.) bulb produces only one or two blooms each year, a low-cost but risky strategy should the flowers be damaged by animals or cold weather.

apportion their reproduction budget in different ways, as already seen in the example of the different flowers of wind- and animal-pollinated species (see Chapters 2 and 3). The costs of pollination to the animal are those associated with any foraging activity and include movement and flight, which is often energy expensive. Reducing the time spent looking for food will therefore reduce the costs to the animal. These reductions will be greater if the animal has a good memory, and they will also be greater if the plant puts out strong, clear signals to attract the pollinator. In this latter case the plant appears to be helping the pollinator, but it does this only because the potential benefit of increased advertising is increased reproductive success. The more accurately targeted and private the advert, the greater the chance of the pollinator remaining loyal to that plant species and hence a reduced risk of the pollinator being distracted. This does not mean that plants with generalised pollination systems cannot be highly successful.

This chapter looks at the various benefits plants provide for their pollinators. The cost to the plant must be acceptable for it to persist with its strategy in the longer term. Some biologists see a conflict here, in that the plant wants to minimise its costs while the animal wants to maximise the benefits that it derives from the plant. This results in plants coming up with ways to prevent visiting animals from overexploiting them. The plants may also invest in defence mechanisms to prevent the wrong animals from gaining benefits. In summary, pollination is associated with a variety of costs and benefits, and these are much easier to identify than they are to measure.

REWARDING LOYALTY
Honeysuckle (*Lonicera periclymenum*) flowers have long, narrow tubes, so only a handful of moths can access the nectar, a strategy that has both benefits and risks.

Lies and deception

At least 7,500 species of flowering plants are rewardless, which is to say that they put out a signal and then fail to fulfil their part of the contract with the pollinator. If bees and other pollinators are so smart, then why do they fall for the deception so often? It would appear that part of the problem is connected with spotting the cheats. If the plant's nectar is scented and that scent is the signal, then there is no problem. However, if the nectary is hidden – perhaps at the bottom of a spur to reduce thieving – then the pollinator does not find out that it is empty until it has looked.

The absence of nectar does not mean that a species is rewardless – it might be that the flower has been visited recently and the nectary has not yet refilled. So, the pollinator cannot assume that it has selected the wrong flower just because there is no nectar. It appears that pollinators have learnt to live with the disappointment.

There is also the argument that it is too dangerous for the pollinator to start rejecting potential food supplies on the basis of a minority of failures. A similar argument is made for the persistence of brood parasites in species like common cuckoos (*Cuculus canorus*) laying eggs in the nests of Eurasian reed warblers (*Acrocephalus scirpaceus*) – in this case it is too dangerous a precedent for the warbler to start kicking out eggs or chicks it does not like the look of.

CREATING A STINK
Flowers of *Stapelia gigantea* are foetid and hairy, and deceive the blowflies that pollinate them by not giving a reward.

Food and water benefits

The biomass of arthropods (insects and their closest relatives) per hectare is nearly a hundred times that of birds or humans, or of mammals excluding humans. However, the average biomass of plants is about a hundred times that of the arthropods. This suggests that most plants are either indigestible or even toxic to herbivores, otherwise there would be none left. That being said, pollination is a relationship that involves plants intentionally making food available to animal pollinators. It is therefore easy to understand why animals would be so grateful when they find what looks like a free meal, especially when that meal includes water as well as energy and other nutrients.

The food and water benefits offered by plants to pollinators come in four forms: nectar, pollen, exudates from the stigmatic surface and floral tissues. It is clear that at least three of these had an original role that was not connected with providing a benefit to an animal visitor. Nectar is possibly the odd one out here, but even nectar may have had another function, because nectaries are found on ferns and these do not produce pollen. These four rewards are examined here in descending order of the frequency in which they are found in nature.

FREE RIDE

Gaping Dutchman's pipe (*Aristolochia ringens*, above) employs deception to enact pollination without providing reward.

Pollen is extremely nutritious, so expensive for plants to produce.

NECTAR SPURS
Western columbine (*Aquilegia formosa*) conceals its nectar in upward-pointing spurs.

At first sight nectar appears to be a very good benefit for both parties in the relationship, being easy for the plant to produce and being used by most pollinators. It is also relatively easy for the animal to take in, either by supping or sucking, and easy for them to store internally. Metabolising the sugars is not difficult for animals and so nectar is an almost immediate fuel supply. However, while it may be a good benefit, it is not perfect. After it is secreted into the nectary, nectar is exposed to the environment. There, it can become more concentrated as the water evaporates, making it more attractive to robbers. Alternatively, in the tropics where humidity is high, it can become more dilute, making it less attractive as a benefit. Because so many animals use nectar as a food supply, thefts are common. Hiding the nectary in a spur is one way to reduce evaporation, reduce dilution by rain and reduce theft.

What is the appeal of nectar?
Although nectar is a quick and convenient supply of energy, it does not offer much else. Even its energy content is not that impressive compared to the energy in pollen: weight for weight, nectar has to be 60 per cent sugar to match the energy content of pollen. Nectar does contain a few amino acids, but these appear to be more associated with the production of scents than benefitting the pollinator with a supply of nitrogen. So what is its appeal to animals?

The 'leaky phloem' hypothesis (see box on page 135) implies that nectar did not develop for the specific purpose of benefitting pollinators, but that it subsequently acquired the function of benefit. It was not even an exaptation, but a 'spandrel' that became useful. However,

Nectar

Nectar is easier to study than pollen and so more work has been carried out investigating its use as a benefit. Nectar production can vary between flowers on the same plant and between plants in the same population, and it can vary with the seasons, with the time of day and with the prevalent environmental conditions. Production can also be affected by the visits of pollinators, or even the *absence* of visits from pollinators. The pattern of production itself can also vary. Nectar may be produced in a fixed volume once in each flower, or it may be topped up following each pollinator visit. It may be produced in an apparently uncontrolled flow for either a fixed period or for the lifetime of the flower. Its production may be linked to pollinator visits or to the stage in the development of the flower. And the pattern may be fixed within a species or the same species may show a variety of patterns.

THIRSTY WORK
The nectar of tower of jewels (*Echium wildpretii*) provides much-needed moisture to pollinators.

several people have suggested that the appeal of nectar is as much the water it contains as the other things dissolved in that water. This is a very appealing idea. For example, although larger bees tend to like nectar with a sugar concentration in excess of 30 per cent, the species that pollinate the tower of jewels (*Echium wildpretii*) in the arid centre of Tenerife – an environment where water is a precious commodity – are satisfied with a 11–12 per cent sugar concentration.

Nectar in non-flowering plants

We know that nectaries exist on the fronds of some ferns in four different families, including the cosmopolitan species bracken (*Pteridium aquilinum*). The most common explanation for these nectaries is that the nectar they contain attracts ants, which will then help themselves to some of the herbivores on the fronds, such as caterpillars. It is interesting to note that some of the nectar is 'stolen' by non-predatory animals, in the same way that nectar is robbed from some flowers. This is similar, but completely unrelated, to the way nectaries on the cyathial cup of spurges (*Euphorbia* spp.) attract ants that then attack other visitors.

A fluid-like nectar is also produced by gymnosperms, although here it is always associated with reproduction and it is not really nectar (which is why some authors refer to it as sexual fluid). Included in these fluids are the pollination drops produced by most extant gymnosperms. The difficulty for biologists arises from the fact that the chemical composition of the pollination drops of gymnosperms and the nectar of flowering plants are very similar. This might indicate that flowering plants evolved from present-day gymnosperms, but we know that is not true and the origin of flowering plants continues to remain a mystery.

Just because the chemical compositions of different substances are similar does not mean that they necessarily have the same function. This is true in the case of pollination drops and nectar. In gymnosperms the fluids are a pollen-capturing mechanism more analogous in function to stigma secretions (see page 33), whereas in angiosperms they are a pollinator-rewarding mechanism. It is still unclear if the pollination drops in fossil gymnosperms were acting as nectar, but in extant species such as cycads, the reward appears to be nectar. In both pollination drops and nectar the fluid contains sugars and some amino acids, in particular proline. They also both contain proteins that appear to have a defensive function against fungal and bacterial spores. It is currently believed that these similarities between the two fluids are just a case of convergent evolution.

NECTAR WITHOUT FLOWERS
Ferns procreate via spores, but some also exude nectar.

NOT NECTAR
Conifer pollination drops catch airborne pollen on the female cones, enabling gymnosperm pollination.

Location of the nectaries
Although nectaries are produced in many different places on flowering plants, the majority are associated with the inside of flowers. Most are found on the non-sexual sepals and petals, with sepals the most likely location. In some flowers, such as buttercups (*Ranunculus* spp.), the nectary is clearly visible at the base of the petals when viewed with a ×10 hand lens. In the closely related hellebores (*Helleborus* spp.), each petal is reduced to a relatively big nectary and the common role of petals in attracting a pollinator by being big and colourful has been reassigned to the sepals.

The nectaries of hellebore flowers (*Helleborus* spp.) are so large that there is sufficient volume of nectar in them to support colonies of yeast. The fermenting yeast gives off heat, which in turn warms up the flower. It is thought that the heat has two potential benefits: it acts as central heating for the flowers, which often have to tolerate snow; and the warmed scent produced by the flowers will travel further and potentially attract more pollinators.

Another delightful explanation for the role of nectar fermentation is that it makes the pollinators drowsy. The alcoholic nectar in some epiphytic orchids growing on the branches of trees is thought to encourage the pollinators to linger for longer, therefore increasing the chance that they will be united with a pollinium (see also page 144).

In the flowers of violets (*Viola* spp.), the nectaries are part of the stamens, whereas in many members of the carrot family (Apiaceae) there is a nectary at the base of the style, just inside the sepals, which themselves are attached to the top of the ovary. Nectary discs are often found at the base of the ovary in flowers where the sepals and petals are attached to the bottom of the ovary. Occasionally, the nectaries are part of the bracts that are often associated with flowers, as in the spurges (*Euphorbia* spp.).

HOT HELLEBORES
The green nectaries at the centre of the flower keep the bloom frost-free by nectar fermentation.

FLOWER CLUSTERS
The fused green bracts surrounding these spurge (*Euphorbia* spp.) flower clusters also bear nectaries.

FREE REWARD
Wasps like this sand-tailed digger (*Cerceris arenaria*) feed on nectar but without providing pollination services.

Nectar production

Nectar is produced in one of two ways, and plants may adopt just one method or both. In the first, sugar is produced during photosynthesis by the plant and accumulated by cells in the nectary. This can be done only during the daylight hours when the plant is actively photosynthesising, and is not an option if the plant flowers before its leaves appear, such as some tropical trees and Mediterranean bulbs. In the second option, the plant makes nectar from stores of starch. This production system can take place at any time of the day or night, and the rate of production can be higher than that found in the nectaries, so it is favoured by plant species that feed greedier pollinators.

THE 'LEAKY PHLOEM' HYPOTHESIS

The evolutionary origin of nectar production is contentious, but it is important to note that the flowers (where most nectar is produced) are not supplied with water by the xylem, as happens in leaves. Instead, flowers are supplied with water by the phloem, and they need lots of it. Scientists have therefore proposed that nectar first appeared in flowers as a 'waste product' left behind when most of the water had been used by the various parts for cell expansion. This is known as the 'leaky phloem' hypothesis and seems to be a reasonable idea in the absence of an alternative explanation.

If nectar essentially originated as a waste material and therefore has a low cost to the plant, this might go some way to explaining why nectar robbers are so common in some situations. These animals either enter the flower in the correct way to access the benefits but do not carry out pollination, or break into the flower and damage it in order to gain access to the benefits. In one study carried out in 2017 on members of the bignonia family (Bignoniaceae), 98 per cent of the flowers were visited by nectar robbers and 75 per cent of the visitors were robbers.

FLOWERS BEFORE FOLIAGE
Sternbergia clusiana daffodils flower before the leaves appear, so any nectar produced must come from starch in the bulbs and not from photosynthesis in the foliage.

The recipe for nectar

Nectar is mostly water plus three sugars: sucrose and its two components, fructose and glucose. Some pollinators (including hummingbirds, hawkmoths and the larger bees) seem to prefer sucrose, while others (for example, bats, perching birds and many insects) prefer glucose and fructose. In addition to water and sugars, nectar also contains some amino acids in varying amounts, with the highest levels of these found in carrion- and dung-mimicking flowers. The amino acid proline is commonly present, but this is a conundrum because it is more important for the developing pollen grain than for the pollinator – so what is it doing in nectar? The answer may be that it stimulates the feeding activity of the pollinator. The colour, taste and scent of pollen can vary, and it is believed that these are important for building a relationship with particular pollinators.

In addition to the mixture of raw ingredients, nectar can vary greatly in its concentration and the volume produced. Not surprisingly, the volume produced is crudely proportional to the size of the pollinator, but it also varies between habitats, between species and even between flowers on the same plant. The lower concentrations (up to 30 per cent) seem to be associated with bats, long-tongued hummingbirds and most moths and butterflies, while higher concentrations (over 30 per cent) seem to be preferred by bees. In plant species with

THE RIGHT RECIPE

In order to attract this sword-billed hummingbird (*Ensifera ensifera*), the flowers of the red angel's trumpet (*Brugmansia sanguinea*) must produce nectar of the preferred sugar type and consistency.

SWEET TREAT

Large bees prefer nectar rich in sucrose.

separate male and female flowers, the latter seem to have nectar with a lower concentration, possibly to persuade the pollinator to make only short visits – one dab of pollen is enough.

It would be presumptuous to assume that a plant brews nectar to a standard recipe. Experiments on stinking hellebore (*Helleborus foetidus*) flowers found a variation in the composition of the nectar, as measured by the proportions of sucrose (0–100 per cent), fructose (40–100 per cent) and glucose (0–40 per cent). There was variation between plants in the population, between flowers on the same plant, and between nectaries in the same flower. The most variation was found between flowers on the same plant, followed by nectaries on the same flower, and the least variation was seen between plants. It is suggested that this variation keeps the pollinators moving swiftly between plants as they seek flowers with high-quality nectar, a strategy presumably aimed at promoting outcrossing.

Pollen

Modern flowering plants overproduce pollen and seeds because both are so attractive to animals. Seeds are a survival capsule containing an embryo and food for the journey from the parent, and so make good fodder. Pollen grains are different in so far as they are much smaller and therefore are unlikely to be present in large enough quantities to be the sole food supply for an animal much bigger than a bee. Because pollen is so small, its surface-to-volume ratio is very high and the pollen coat makes up at least 30 per cent of the grain. Since the function of this coat is protection of the few cells inside the grain, it is very difficult to penetrate and very difficult to digest. Beetles seem to be masters at cracking open pollen, while thrips suck out the contents. In many cases the contents of the grain are sucked out by the pollinator after the pollen has entered its gut, sometimes through the aperture intended for the pollen tube. Some pollinators even wait for the pollen to germinate in their gut before digesting the contents. Bees and pollen wasps (subfamily Masarinae) seem to have cornered the market in pollen digestion, with research indicating that they can extract up to 83 per cent of the protein content of the grains.

Pollen provides a more balanced diet than nectar, although the proportions vary between species. Its protein content can be anything from 2.5 per cent to 61 per cent, starch content 0–22 per cent, lipid content 11–18 per cent and energy content 16–28 J/g. The variations appear to have very little to do with how the plant is pollinated and more to do with what its ancestors put in their pollen. This is a biological example of the philosophy 'we've always done it this way and don't see why we should change it now'. It could also indicate that it is very difficult to change pollen without changing many other things at the same time. Pollen does contain significant amounts of phosphorous and other useful minerals, but its capacity to supply nitrogen is very important. In fact, pollen can be almost as rich in nitrogen as animal tissue, whereas seeds are never quite a match for this. Perhaps we should be spreading pollen on our buttered toast in the morning rather than honey,

POLLEN PREDATORS

Some pollinators prefer pollen over nectar, such as these rhinoceros beetles on daylily (*Hemerocallis* sp.)

Is this co-adaptation?

The lifetime reproductive success of bees and bee colonies appears to be strongly affected by the nutritional quality of the pollen the insects collect, and so this may be a good example of co-adaptation.

Another striking example comes from the genus *Agave*, where different species use a range of different pollinators. In those species pollinated by bats, the pollen protein content is as high as 43 per cent, whereas in species pollinated by insects it is 8–10 per cent. Another interesting observation is that the pollen of the giant saguaro cactus (*Carnegiea gigantea*) is exceptionally rich in two amino acids, tyrosine and proline. The former is important for the manufacture of the collagen that makes up the membrane of the bats' wings – the collagen is 80 per cent tyrosine. The proline is an important component of the bats' milk, but it is also very important for the formation of the pollen tube. As the style of the cactus is very long, the proline may be present primarily for the benefit of the pollen and the bat is just lucky.

CACTUS CO-EVOLUTION

Saguaro cactus (*Carnegiea gigantea*) blooms open primarily at night when the lesser long-nosed bat (*Leptonycteris yerbabuenae*) is active.

Nectarless, pollen-only flowers

Many plants do not produce nectar. While it would be tempting to assume that these are all wind pollinated, about 20,000 species produce only pollen and they are pollinated predominately by bees. These bees tend to use the buzz collection method of gathering pollen (see page 91), which is a neat trick that allows them to keep the pollen to themselves and at the same time allows the plant to restrict its benefaction.

Included in these pollen-only species are poppies (*Papaver* spp.), rockroses (*Cistus* and *Helianthemum* spp.), cinquefoils (*Potentilla* spp.), St John's worts (*Hypericum* spp.) and meadow-rues (*Thalictrum* spp.). The plants often have very conspicuous anthers, which is to be expected, and some (including meadow-rues) are also wind pollinated. It has to be noted that this system works only if the flowers are bisexual, because the pollinator would not visit a female flower. The evidence appears to support the hypothesis that nectarless, pollen-only flowers produce more pollen per ovule than plants that also produce nectar.

Stigmatic exudates

The fact that plants produce substances from the stigma is not unexpected. The functions of these exudates include helping to capture the pollen and sticking it to the stigmatic surface for long enough such that the same liquids aid with the germination of the pollen. However, as soon as a plant produces anything, some animal will come along and examine it, lick it and/or eat it, and if it provides nutrition and does not kill the animal, a relationship can develop. Stigmatic exudate is mostly made up of lipids and amino acids, and so is useful to animals. The animals trapped in the floral dungeons of dutchman's pipe plants (*Aristolochia* spp.) are sometimes sustained by the stigmatic exudates of the female flowers until their release.

The common spotted orchid (*Dactylorhiza fuchsii*) belongs to a genus that has a reputation for deceiving honeybees and bumblebees, but it actually does give something back in the form of the oily exudate rich in glucose and amino acids. One common feature of flowers that hand over stigma exudates is that the stigma has to be quite robust to avoid being damaged. This is not a problem in orchids, as the stigmatic surface is a part of the column, which is generally a robust structure.

CHOCOLATE REWARDS
The protruding purple staminodes of cacao (*Theobroma cacao*) provide food for pollinating midges.

Assorted floral parts

Herbivores eat plants and any part will do. The consumption of entire flowers is not going to end well for the plant, but if the herbivore eats just part of the flower things may not be too bad. Some flowers do have food bodies, containing sugars and starch, lipids and proteins. Beetles (pollinating *Calycanthus* species while eating the swollen tips of the tepals), bats (pollinating *Freycinetia* species while eating the fleshy flower bracts) and non-flying mammals are the major consumers of these freebies. Perhaps the best-studied example is the giant Amazon water lily (*Victoria amazonica*; see pages 148–9), but the most important are undoubtedly the staminodes in the flowers of cacao tree (*Theobroma cacao*), which are pierced and sucked out by the pollinating midges.

Non-food benefits

Pollinators have needs other than just nutrition and fluids. Some plants have evolved means of satiating some of the very specific requirements of a relatively small number of important pollinators.

Oils

Oils, particularly aromatic oils, occur throughout the plant kingdom and the list of uses to which plants put them is well into double figures. About 1 per cent of plants synthesise predominately scentless oils that end up being exploited as a benefit by pollinators, and the majority of these do not synthesise nectar. The role of these oils in pollination was only really taken seriously following the pioneering work of Stefan Vogel in the 1960s and 1970s. Since then they have become recognised as possibly the third most common form of bribe for pollinators after nectar and pollen.

FILLING UP

In southern Africa, twinspur (*Diascia barberae*) flowers use oils to reward pollinating *Rediviva* bees. The flowers have two rear-facing spurs that contain trichomes, hair-like structures with black tips, which secrete the oils. *Rediviva* bees have long, slightly hairy front legs that reach into the spurs and harvest the oil.

The taxonomic spread of plant species that produce oils tends to be concentrated in a few families, but these families are spread widely across the evolutionary tree. The oils are produced in structures called elaiophores, which come in two types, trichome and epithelial. Trichome elaiophores are essentially glandular hairs that secrete oils onto the surface of the flower. They are most commonly found on petals but may be seen on stamens or ovaries. Trichome elaiophores are found in several plant groups, including *Bactris* palms, whose flowers are pollinated by beetles while they eat the hairs and mate – multitasking at its most excessive.

Epithelial elaiophores consist of an oil-filled pustule, which the pollinating bees have to rupture. It has been speculated that this type of elaiophore is a modified nectary. The most exaggerated examples are found in the taxonomically isolated, parasitic woody genus *Krameria*, where two of the five petals have become modified into oil glands up to 4 mm (0.16 in) in diameter that are slightly reminiscent of the nectaries of hellebores. Each petal can contain 0.5 ml (0.02 fl oz) of oil.

The oils help to protect the pollen from desiccation, and it is possible that they also have value to the plant as an adhesive – pollen mixed with the oil is more likely to stick to a stigma. The oil molecules are long-chain carbon molecules (14C–18C), meaning that they are much bigger than the scents described Chapter 4. The oils are generally colourless or pale yellow and are odourless, so the plants rely on their shininess to provide a visual signal. Almost inevitably, some mimic orchids produce shiny growth that is mistaken for drops of oil by unwary bees. In at least one oily species, dotted loosestrife (*Lysimachia punctata*), the flower gives off a unique scent thanks to an unusual ketone, which means that it is a very reliable signal.

Looking at the animal evolutionary tree, oil foraging has evolved separately at least six times in bees. The bees use the oil to waterproof both their entire nest and the individual cells. Furthermore, the larvae may consume the oils along with pollen collected by the bees. Many of these bees use buzz collection, and so the plant and bee appear to be working in partnership to restrict access to the pollen.

OIL SPOT

Hungry pollinators look for the gloss of oily rewards, but dotted loosestrife (*Lysimachia punctata*) also provides a unique fragrance.

The solitary bees that collect oils have modified limbs called elaiospathes, which they use to rupture the elaiophores. Other modified limbs act as mops to collect the oils running out of the elaiophores. Oil-collecting bees generally visit many plants, whereas the plant species tend to have just one pollinator. This is another example of how pollinators cope with the fact that their needs extend over a greater period of time than any one plant species is in flower.

PETAL PRODUCTION

Each white rhatany (*Krameria bicolor*) flower has two small modified petals that act as oil reservoirs.

PRETTY IN PINK
Two large bracts advertise the rather small flowers of *Dalechampia dioscoreifolia*.

Resins and gums

Many plants produce gums – species in the Australian genera *Eucalyptus*, *Angophora* and *Corymbia* are actually commonly known as gum trees. The function of the gums, if there is one, may be as a defence against infections. A small, rather boutique pollinator–plant relationship exists between a few plants, such as balsam fig (*Clusia rosea*) and *Dalechampia* species, and a few bees from just two families, including euglossine bees. In these relationships resin is collected by the bees either for constructing nests, or for internal structures such as cells. The resin is particularly well suited to the latter because it comes complete with antibiotics that help reduce fungal and bacterial infections in the larvae.

The nests that are made using the resin are exquisite little constructions reminiscent of the nests made by Eurasian wrens (*Troglodytes troglodytes*), except that the bees' nests have doors that can be closed. The resin is used in one of two ways: to glue together pebbles in a way reminiscent of pebble-dash houses, or to stick together strips of bark. Some parts of the nest do not harden and stay malleable.

Scent

Scheming orchids

The use of scent as a benefit is almost exclusively seen in relationships between euglossine bees and orchids, with about 200 bee species involved. The trickery used by orchids to lure in the pollinators has led to the evolution of some bizarre structures.

The female bees are not involved and feed in the normal way on nectar and pollen. The male bees, on the other hand, are solitary and live for between three and six months, which is a relatively long time for a bee. They feed on nectar (if available) and scrape up the scents that are produced in special glands on the orchid's labellum petal.

The male bees have specially modified limbs that can mop up the scent, which is then stuffed into the equivalent of perfume bottles on their legs, each of which can hold up to 60 µl of fluid. Each flower produces a mixture of six to ten different compounds similar to floral fragrances, but the bees visit more than one species of orchid and mix up their own species-specific fragrant bouquet. This means that tight one-to-one species relationships do not evolve. The scents are so powerful that biologists studying the bees can easily entice them with artificially synthesised compounds.

The male bees trap-line the orchids but after every few flowers they have a rest, and at this point they remove some of the scent and waft it around. It has always been assumed that this behaviour is to attract females, and possibly to warn off any males nearby. However, there is little direct evidence that this is actually the case. That said, given that the males mate only once in their lives, they do have a strong incentive to attract the correct mate and to mate successfully.

For what is a relatively small group of relationships between the orchids and the bees, there is a surprisingly wide variation in the strategies played out. However, one thing they all seem to have in common is that collecting the scent takes a long time and is very different from the grab-and-go strategy of so many other pollinators. During the process the bees often become sluggish and seemingly intoxicated, which gives the plant an opportunity to control the bee accurately and specifically.

NEST MATERIAL
The flower clusters of *Dalechampia dioscoreifolia* include three female flowers (front, red), two male flowers (behind, with yellowish pollen) and a resin gland (back). Because bees rely on the resin to build nests, the plant has obligate pollinators.

In the epiphytic Central and South American *Stanhopea* orchids the structure of the flower is such that the male bee has to cling on with its hind legs and hang its head downwards to collect the scent (not dissimilar to kissing the Blarney Stone). Some bees inevitably lose their grip and fall onto the sticky pads that are attached to the pollinia. In some species the orchid even provides a chute down which the bee slides; this orientates the bees more accurately, presenting its body in the right way to the sticky pads.

The orchid genus *Catasetum*, also from Central and South America, is odd even for orchids in that its members have unisexual flowers that are grouped in separate inflorescences on the same plant. As the male bee forages around on the column looking for the scents, it disturbs a trigger and the pollinia are forcibly shot onto its body. When the male bee visits a female flower, it has to hang upside down as described above. However, this time a pollinium (which is attached to an elasticated thread) swings down under the influence of gravity in an arc and collides with the stigmatic surface. Each *Catasetum* species was thought to emit a different bouquet and thus be pollinated by a different species of bee, but scientists are now not so sure. The species often place their pollinia on different places on the bee's body, and there is a report of one bee carrying 11 pollinia from 11 different orchid species.

The iconic species in this little group of scheming orchids are the bucket orchids. In these, the labellum is distended into a bucket about 20 mm (0.8 in) in diameter, and overhanging this is a pair of protruding scent glands from which the bee sups. Eventually the bee falls intoxicated into the bucket, and as the sides are slippery and steep it cannot climb or fly out – the only escape is through an emergency exit in the side of the bucket. As the bee crawls through the tunnel, any pollinia on its back are scraped off. Further down the tunnel is another pair of pollinia for the bee to take away to the next flower. The bee is prevented from returning to the scent glands over the same bucket because the flower shuts down the production of scent – a mechanism to avoid selfing.

ON THE HEAD

Orchid pollen is clustered into pollinia and different orchid species attach their pollinia to different parts of their pollinator.

BUCKET ORCHID
Coryanthes macrantha.

MALE FLOWERS
Catasetum osculatum.

An arum enigma

Still on the subject of scent as a benefit, a 2017 report suggested that *Anthurium acutifolium* in the arum family (Araceae) is pollinated by a male oil bee, *Paratetrapedia chocoensis*, which collects scent off the spadix. The spadix lacks the elaiophores the bee is looking for, and as the bee rummages up and down, he either picks up pollen (if the plant is in the male phase) or deposits pollen from another plant (if it is in its female phase). What the female bees are doing at this time is not recorded, but they may be pollinating a different species. This is possibly an example of a species transitioning from one pollinator to another. This must happen often, because plant species appear to evolve and speciate and different rates, and so sometimes a plant will have to recruit a replacement pollinator if it is to avoid becoming extinct.

BUCKET LIST
Having harnessed a bee to carry its pollinia, this *Coryanthes panamensis* bucket orchid allows the insect to escape.

Heat: a controversial benefit

It is very easy to anthropomorphise biology, and the idea that a warm flower will be more attractive than a cold one is a very appealing idea. As seen on page 134, the activity of yeast in the large nectaries of hellebores can generate heat, and in 2017 it was reported that warm flowers of the stinking hellebore (which often appear when there is snow on the ground) are more likely to be pollinated than cold flowers.

More than 900 plant species have now been recorded as having thermogenic flowers, most of them tropical and most pollinated by beetles. However, there has been a reluctance among some pollination biologists to accept that providing heat could actually be considered a part of the bribe to, or reward for, visiting a flower. Their argument is that the heat is just there to make the scent travel further, thus making it part of the advertisement. However, the heat continues after scent production has ceased, suggesting that the control of heat production is not linked to scent production, and some plants continue to produce heat long after the pollinator has arrived. The most compelling example of this is the poster plant for the plant kingdom, the giant water lily (see pages 148–9).

The French Guianan *Philodendron solimoesense* is pollinated by the same scarab beetles (*Cyclocephala* spp.) that pollinate the giant water lily. Again, the heating is left on long after it is needed to make the flora odours travel further. Given that this plant lives at lower temperatures than the giant water lily, the benefit to the beetles of the heat it provides will be even greater.

However, this does not mean that every warm flower is going to provide accommodation. In some thermogenic species the heat is simply produced to maximise scent dispersal. Similarly, a flower does not have to provide heating in order to be a good bed for the night. In the UK, it is a common sight in gardens to see buff-tailed bumblebees (*Bombus terrestris*) asleep on the flowers of the plume thistle (*Cirsium rivulare*). In the south of Portugal, meanwhile, early-rising biologists can see bees emerging from an overnight stay in the flowers of the tongue orchid (*Serapias lingua*). The bees can stay 3° C (5° F) warmer in these sleeping bags than if they slept unprotected under the stars – and without any heat production on the part of the plant.

The generation of heat is a tangible and measurable cost to a plant, and if Charles Darwin was right then it must therefore also provide a benefit to the plant. There are many ideas in the scientific literature about the potential functions for the heat in addition to those already discussed, including increasing the biochemical activity of the enzymes involved in scent production, being a by-product of the production of carbon dioxide used as a signal or an anaesthetic, or as an aphrodisiac. It is not unusual for a plant trait to have many different uses in different plants, and thermogenesis is no exception.

HOT TOPIC
Like many other aroids, *Philodendron martianum* can heat its flowers up.

Brood chambers and nursery provision

There is one more benefit that plants supply to animal pollinators, and for some people it is the most fascinating. The relationships between figs (*Ficus* spp.) and fig wasps (superfamily Chalcidoidea), and between yuccas (*Yucca* spp.) and yucca moths (family Prodoxidae) have been described in detail many times, but no book on pollination would be complete without at least a brief description of these and some of the other 14 known nursery or brood-site mutualisms.

There are three types of brood-site relationships. The best known is where the pollinator's larvae parasitise the ovules and thus the seeds. Yuccas, figs, woodland stars (*Lithophragma* spp.), globeflowers (*Trollius* spp.) and *Pachycereus* cacti use this system. Second, there are the plants where the larvae feed on the pollen but do not take it away to another plant. The insects involved here are all thrips and can be very destructive of crops. Finally, there are those plants where the larvae grow on the old flower after pollination has taken place. Included here are some cycads, palms, breadfruit (*Artocarpus altilis*) and, perhaps best known, the dutchman's pipe plants (*Aristolochia* spp.).

PLANT NURSERY
Seed consumption by pollinator larvae can reduce seed set to 20 per cent in some globeflower (*Trollius* spp.) populations.

The giant water lily

The giant water lily (*Victoria amazonica*) is probably best known for its remarkably strong leaves, which can support the 23 kg (51 lb) weight of a prone child. It is less well known for its flowers, partly because they open at about 6pm when most botanical gardens are closing their gates.

MASSIVE LEAVES
The two species of *Victoria* water lilies, *V. amazonica* and *V. cruziana*, both hail from South America, where they dominate ponds and river backwaters.

The work of botanist Ghillean Prance and entomologist Jorge Arias in 1975 on the species in its native habitat in the Amazon, along with that of Roger Seymour and Philip Matthews from Adelaide 31 years later, has given us a very clear description of the pollination of this extraordinary plant.

Day one

The giant water lily's flowers last just two days. On the evening of day one, the flowers have white petals and emit a strong fruity odour, which the heat in the flower helps on its way. Passing scarab beetles in the genus *Cyclocephala* enter the flower down a narrow vertical tunnel, the walls of which consist of the stamens tightly pressed together. The beetle finds itself inside the floral chamber, at the centre of which is the female part of the flower. It is important at this point to understand that the flower is protogynous, meaning that at this time the female stigma is receptive to pollen but that the anthers on the stamen are tightly shut. Hopefully, the beetle has brought some pollen from a flower on another plant (it is very unlikely to be from a flower on the same plant because they rarely produce more than one flower at a time).

The beetle does not leave the floral chamber – and why would it? The chamber is warm, there is a plentiful supply of food and if the beetle is lucky it will find a mate for the night. The warmth is supplied by the cells in a ring of protruding 'stylar processes' around the top of the chamber, which are full of starch as well as being warm. The temperature inside the chamber can be up to 9.5° C (17° F) warmer than that outside, which is important to the beetles as they function best at a body temperature of 31° C (88° F). The centrally heated chamber saves them up to 80 per cent on their metabolic fuel bill. Having stayed open long enough to admit several beetles, the tunnel of anthers closes and the insects are trapped. The production of scent is turned off at this point but the heating remains on.

PURE WHITE
A strong fragrance first attracts pollinating beetles to the white blooms of *Victoria amazonica*.

Day two

During day two, the petals start to turn pink, becoming a deeper cerise as evening approaches. The generation of heat in the flower moves from the stylar processes, some of which have been eaten anyway, to the stamens. The heat they generate is not as great as that provided the previous evening, but it is enough to keep the beetles at 'flight temperature'. As the petals turn to cerise they begin to peel back, revealing the stamens, which then peel back too, thus opening the sealed tunnel and allowing the beetles to escape. But the beetles are now coated in pollen, because on the evening of day two the anthers open. The insects emerge into the cooling evening air and smell a familiar strong fruity odour coming from a welcoming white flower nearby. Well, what is a beetle to do, but to enter its chamber and settle down for a warm night?

This is very definitely pollination with benefits. It has been shown that these lucky beetles spend most of their lives inside the giant water lily flowers, and that the heat from the flowers saves them a great deal of energy. There can be no doubt that this bed-and-breakfast package is a serious reward and that the thermogenesis is a very important part of that reward.

IN THE PINK

As the petals peel back, pollen-coated beetles are released from their floral prison, ready to find another flower (right).

Although white initially, the flowers turn pink on day two and stop producing scent, so beetles know not to visit (below).

The rules of engagement

One feature all the species involved share is that their very specific relationships have to be re-established every generation. There are no alternative suppliers of a place in which the insects can lay their eggs, and the signal from the plant therefore has to be strong and unambiguous. Work published in 2010 revealed that we still have a lot to learn about how this is achieved, but it has been suggested that scents – particularly volatile organic compounds – are very important, along with some visual signals. From the relationships investigated so far, these signals involve mixtures of just a few specific compounds and appear to be private channels of communication. This implies that the simpler the signal, the less easy it is for either party to break out of the relationship. The signal has to indicate not only the location of the flower, but also that it is at the correct stage of development.

MOTH MANSION
Senita moth (*Upiga virescens*) females actively pollinate their namesake cactus, but not only for nectar. The plant also provides a place for their larvae to pupate and all the food they need.

There is another interesting point to highlight about these very close, mutually beneficial relationships. This is that they can be stable over the longer term only if there is a mechanism on both sides for preventing overexploitation – in other words, they both have to play nicely. Since the yuccas and their moths appear to have been in cahoots for perhaps 40 million years, since the Eocene epoch, it would be fair to consider them stable relationships.

The senita cactus

The relationship between the senita cactus (*Pachycereus schottii*) and the senita moth (*Upiga virescens*) was described for the first time in 1998 by Theodore Fleming and Nathaniel Holland. The senita cactus is one of the iconic cacti of the Sonoran Desert in North America, with swollen, grooved, spiny green stems. The flowers open at night, when perhaps water loss will be reduced. Female senita moths actively collect the pollen on modified scales on the sides of their abdomen. The female takes the pollen to another flower and deliberately places it on the stigma by rubbing her abdomen on the structure for up to 20 seconds. She then lays an egg on the tip of a petal of the flower, before crawling down into the centre of the flower for a drink of nectar. Male moths also visit the flowers, but they just partake of the nectar and have never been found carrying pollen.

After their night-time opening, the flowers close and the moth eggs develop within. The larva that hatches from an egg crawls down to the ovary, where it consumes some of the young seeds. When the larva is satiated, it eats its way down the stalk of the fruit into the stem of

PROS AND CONS
Senita cacti (*Pachycereus schottii*) utilise a moth for pollination, but its hungry larvae destroy some of their seeds.

the cactus, where it pupates. Sadly, this causes the immature fruit to fall from the plant and all the seeds are lost. The relationship survives because only 20 per cent of the larvae survive and they succeed in ruining only 30 per cent of the fruits. It is also important to remember that these are long-lived plants and so can tolerate the loss of some seed each year. In addition, the relationship survives because the plant is self-incompatible, and so the moth is forced to take the pollen to another plant of the same species if it wants to breed successfully.

It has been discovered that in cooler years up to 25 per cent of successful pollinations in senita cacti are the result of visits from two species of bee, so the plant is not as faithful as first thought and the moths are not totally reliable. The presence of a 'substitute' pollinator is undoubtedly an evolutionarily sound strategy.

The yuccas and their moths

The relationship between yuccas and yucca moths is similar to that seen in the senita cactus and its moth, but there are some important differences. For a start, there are more than 40 species of yuccas. Two of them (Spanish bayonet, *Yucca whipplei*, and the Joshua tree, *Y. brevifolia*) have their own unique moth pollinators (*Tegeticula maculata* and *T. synthetica*, respectively) while all the rest have to share the yucca moth (*T. yuccasella*). The relationship between Spanish bayonet and *T. maculata* is the best studied to date. The next difference is that the female moths do not take any of the nectar at the base of the flower, and it is suggested that this is a ruse to keep all other insects away from the stigma.

The female moth deliberately collects pollen from up to four of the six anthers on one flower. She does this by climbing up the stamen and holding herself in place with her tongue while she rolls the pollen up into a ball under her head. She then flies off to another flower, ignoring the stigma on the flower she is currently harvesting. At the new flower, she examines the ovary to check that no other moth has got there first. If there are no signs of previously laid eggs, the female will lay one egg and then climb up the style. At the top, she scrapes some pollen off the ball and slaps it onto the stigma, which form a funnel-like vessel. Next, she then descends the style and lays another egg, and then back up the style she goes. It is unlikely that the female will lay an egg in every ovule, and so some seeds will be left to produce the next generation of yuccas.

Abnormal growth of the ovary provides food for the larvae. The fruits fall to the ground with mature larvae and some seeds (in some species up to 45 per cent of the seeds can be eaten by the larvae). The larvae then burrow into the ground and pupate. They emerge over the following three years, which is fortunate because the yuccas do not flower reliably every year.

MATERNAL INSTINCTS
Female yucca moths (*Tegeticula yuccasella*) deposit pollen, then insert their eggs into the flower's ovaries.

The figs

The flowers in figs are hidden from sight inside the syconium, which develops into the familiar fig 'fruit' seen on greengrocers' shelves. The female fig wasp is attracted by scent to the ostiole, the opening in the syconium, and pushes her way in, sometimes losing her wings in the process. In this way the syconium 'selects' the correct wasp by excluding the wrong species. Inside the syconium are simple female flowers, some of which have short styles and some of which have long styles. The wasp tries to lay an egg in every ovary, but she can only reach the ovaries of the female flowers with short styles. The flowers with long styles are pollinated.

The eggs that are successfully laid develop into male or female wasps. The males are wingless, almost legless and nearly blind, but all they have to do is eat their way into an ovary containing a female wasp and mate with her before they die. The females crawl out of the ovary wall through the hole made by the male. By now, male flowers have developed around the ostiole, and as the female emerges into the light, she takes away some pollen, hopefully pollinating flowers on a different fig tree. The fig tree–fig wasp relationship differs from that between yuccas and the yucca moth in that the female removing the pollen is not the same female that laid the eggs.

Because there are over 850 fig species, most of which have a relationship with a particular species of wasp, this has become a favourite area for biologists to study when investigating evolution and its many forms. Chapter 6 looks at how biologists have used the study of pollination to provide answers to some important questions.

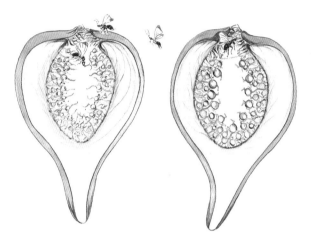

ABOVE: NURSERY FIG

Female fig wasps bring pollen to a syconium (left), fertilizing some flowers and laying eggs in others. Male wasps emerge from the eggs, locate female larvae within the syconium, then breed with them. Newly emerged female wasps pick up pollen from male flowers as they exit the syconium (right).

LEFT: POLLINATION PEST

Figs of *Ficus hispida* are pollinated by the fig wasp *Ceratosolen solmsi*, but a *parasitoid* wasp called *Apocrypta bakeri* lurks nearby. It penetrates the fig with its *ovipositor*, injecting eggs into the developing wasp larvae and reducing the chances of pollination.

Chapter summary

There are two basic categories of rewards or benefits provided by plants to pollinators: nutritional and non-nutritional (although it is possible for a benefit to straddle both). Included under nutritional benefits are nectar, pollen, oils, floral parts and secretions from the stigma. Providing food for larvae that develop from eggs laid in flowers also comes under this heading. Non-nutritional benefits for animals involved in pollination include resins and gums for nest construction, lipids for waterproofing nests, fragrances (perhaps for sexual attraction), mating arenas, heat and warmth, a place to sleep and somewhere to lay eggs.

It is interesting and important to note here that plants cannot use all of these benefits to manipulate and cajole every pollinator. In the case of bees, for example, the two sexes have different ideas on what makes a good incentive. Only female bees will be impressed with oils and resins because they make the nests. Only male bees collect scents to attract females, and only male bees will be fooled into pseudo-copulation. Both sexes will collect and consume nectar, but only females will visit pollen-only flowers. It is also important to note that a pollinator may have different needs that are supplied by different plants. Many plants might supply nectar, perhaps fewer can supply sufficient pollen, and fewer still will be able to satisfy the pollinator's need for oils, resins and fragrances. However, pollination is a system with many interspecific connections that may well be more complex than we imagine.

Chapter 6

The Biological Significance of Pollination

According to the *Oxford English Dictionary*, biology is 'the scientific study of life, dealing with the morphology, physiology, anatomy, behaviour, origin, and distribution of living organisms'. There is no mention of evolution here, which is interesting because according to NASA, life is a self-sustaining system that must be capable of evolving as a result of natural selection. There is no doubt that pollination is an area of biology that can be used to shed light on many aspects of life on Earth and how living systems function, and on how species survive and evolve.

Painted lady (*Vanessa cardui*) on purpletop vervain (*Verbena bonariensis*).

The central importance of evolution

The roll call of biologists who have studied pollination is long and illustrious. Joseph Kölreuter (1733–1806), Christian Sprengel (1750–1816) and Charles Darwin are often cited as the triumvirate who, over a period of about a hundred years, set the agenda for the study of how plants move their pollen from one flower to another. The way we look at life on Earth changed in 1859 with the publication of Darwin's *On the Origin of Species*. From thenceforth, many biologists saw describing and explaining evolution as the central purpose of their work. In 1973, evolutionary biologist Theodosius Dobzhansky (1900–1975) wrote his seminal paper 'Nothing in biology makes sense except in the light of evolution'. At the end of the paper Dobzhansky quotes philosopher Pierre Teilhard de Chardin (1881–1955), who stated that evolution 'is a general postulate to which all theories, all hypotheses, all systems must henceforward bow and which they must satisfy in order to be thinkable and true. Evolution is a light which illuminates all facts, a trajectory which all lines of thought must follow – this is what evolution is.'

Many biologists and natural historians try to explain what happened in the past, while others try to explain what is happening now and what will happen in the future. Not surprisingly, opinions in natural history and biology vary as much as they do in the study of history. Passions can run deep when new ideas suggest that widely held views are wide of the mark, and toys can be thrown out of academic prams. Suggesting that Darwin might have been wrong should not be undertaken lightly. For example, in *On the Origin of Species* Darwin wrote that he 'can understand how a flower and a bee might slowly become, either simultaneously or one after the other, modified and adapted in the most perfect manner to each other, by the continued preservation of individuals presenting mutual and slightly favourable deviations of structure'. Whilst this might be true some of the time, it is not true all of the time.

JOSEPH KÖLREUTER
(1733–1806), German botanist.

CHRISTIAN SPRENGEL
(1750–1816), German naturalist.

THEODOSIUS DOBZHANSKY
(1900–1975),
American geneticist.

PIERRE TEILHARD DE CHARDIN
(1881–1955),
French philosopher.

The biological species concept

It is worth pointing out that Darwin was completely ignorant of the genetic mechanisms that underpin his theory. That had to wait until the twentieth century, long after his death. The melding of genetics and natural selection is sometimes referred to as the new synthesis, and it brought people's attention back to what constitutes a species. In 1942, biologist Ernst Mayr (1904–2005) crystallised the biological species concept, which states that a species is a group of individuals that can freely interbreed to produce fertile offspring that resemble their parents. This works just fine for animals, but not for plants and certainly not for bacteria.

At the heart of the biological species is the requirement for reproductive isolation. This is the simple idea that if a population of similar organisms is split in half and each half can no longer share genes with the other, then they can theoretically go off down their own evolutionary branches. Over time, the two populations will look different. This may be due to the natural selection of different characteristics, or it might be due to a more random drift apart. The jury is out on which of these is more important; as in all biology, it probably depends on the circumstances. If reproductive isolation is the primary requirement for speciation to occur, then pollination can be perceived as one obvious way in which two populations could be kept apart.

ERNST MAYR
The German-American biologist (1904–2005) was a key proponent of the biological species concept.

TIMELY TOME
On the Origin of Species (1859) by Charles Darwin.

The plant species problem and despeciation

The biological species concept came under serious attack when Peter Raven and Susan Eichhorn pointed out in 1966 that reproductive isolation does not appear to lead automatically to new species. They showed that within what we choose to call a species, there are many populations that never mingle their genes and yet continue to look the same, and that when tested can still freely interbreed. Plants further undermine the biological species concept when two clearly distinct species are brought together in cultivation and then promiscuously produce fertile offspring that are completely intermediate in form between their parents (see box).

The birth of a new species?

A simple, well-documented example of two distinct species producing fertile offspring that are intermediate in form happened in Oxford in the 1990s when honey spurge (*Euphorbia mellifera*) from Tenerife was grown a few metres from *E. stygiana* from the Azores.

In nature, the two species live 1,500 km (900 miles) apart on islands in the Atlantic that have very different climates. They do not interbreed, because there is no mechanism for the transfer of pollen. Spurges are pollinated by a range of different animals, from birds to flies, and while some are pollinated by just a few species, others welcome all comers. The two species in question fall in the latter camp.

The fertile hybrids produced, named *Euphorbia × pasteurii* after the undergraduate who proved their parentage, arose at Oxford spontaneously after the pollen was taken from honey spurge to *E. stygiana* by an anonymous peripatetic animal. What this shows is that a shared pollinator can unite a pair of species to create a hybrid species. The fact that the hybrids were fertile with both of their parents, as well as with each other, means that the distinctions between the parent species and the hybrids may become completely eliminated over time,

HONEY SPURGE
Euphorbia mellifera.

AZORES SPURGE
Euphorbia stygiana.

resulting in just one species, not three – in other words, despeciation will occur. It is well known that many plant species are the product of hybridisation, and so this scenario played out in Oxford might be a paradigm for some speciation in the wild. It also provoked the suggestion that if pollinators can undo speciation in Oxfordshire, they can just as easily facilitate the origin of species in the wild.

PASTEUR SPURGE
The result of an accidental cross in a botanic garden, *Euphorbia × pasteurii* exists only because of the actions of a pollinator.

By the time the genetic basis of evolution was unravelled, Darwin had already set running the idea that the relationship between plant and pollinator is mutually beneficial and that the pair will become a better and better fit over time – a concept that was erroneously described in the past as co-evolution. There was also the belief that if flowers and their pollen vectors fit so nicely together, then it should be possible to look at a flower and deduce the identity of the pollinator. The linking of a group of floral characteristics to a species or group of pollinators resulted in the description of pollination syndromes (see pages 165–79).

IMPOSSIBLE CROSS

Digitalis × valinii occurs only in cultivation; its parent species are isolated from one another in the Canary Islands (*D. canariensis*) and Europe (*D. purpurea*).

Can pollinators facilitate speciation?

SORGHUM
Sorghum bicolor is a C4 grass.

The debates on co-evolution led to the idea that if a change arises in the structure of a flower, the flower's pollinator will no longer fit as well and another animal will take over. As explained above, the mutated plant would then become reproductively isolated and able to go off down its own evolutionary pathway, eventually leading to the emergence of a new species. Can the behaviour of a pollinator facilitate and drive the emergence of a new species? If it can, then is there a relationship between the number of insect species and the number of flowering plants?

Many reference books and research papers still state that the diversity of flowering plants is directly influenced by the diversity of insects. However, work published over the past 30 years clearly shows that the relationship between the diversity of insects and the diversity of flowering plants is much more complicated than previously thought. There are many aspects of the angiosperms that have had a major effect on speciation, and most of these have nothing to do with pollination.

For a start, many are herbs, or non-woody. This is in contrast to their nearest relatives, the gymnosperms, which are all woody. Herbs grow faster than woody plants and have shorter life cycles, which means that they can change (evolve) more quickly and thus diverge more quickly. Shorter life cycles have another advantage, in that the plant can survive in a habitat where suitable growing conditions occur for only the minority of the year. Yet another advantage of herbs is that they can be smaller and therefore exist in habitats that are unavailable to large trees. Small annual herbs live fast and die fast, but they also evolve fast.

A second reason for the larger than expected number of angiosperm species is that their leaves can be adapted into a wide range of structures other than flat photosynthesising organs. Leaves can capture and digest insects, they can cling to other plants for support, or they can become hard and sharp to protect the plant. This adaptability helps flowering plants to invade habitats unavailable to other plants. Furthermore, unseen to our eyes, the structures inside photosynthesising leaves can perform biochemical wizardry that enables them to live in hot, dry conditions. Known as C4 (making a four-carbon compound outside the Calvin–Benson cycle) and CAM (crassulacean acid metabolism), these options are not available to any group of plants other than the flowering plants.

PINEAPPLE
Ananas comosus utilises CAM photosynthesis.

Finally, flowering plants also produce a diversity of fruits and seeds. This has further enabled them to disperse to new habitats and to survive periods of inhospitable conditions, including hiding from the effects of wildfires until the rains come again. The way in which the food supply for an angiosperm embryo is produced differs from that found in all but one of the gymnosperms, and this too is thought to contribute to the diversification of the flowering plants.

Fruits develop from the carpel, which is the female part of the flower consisting of the stigma, style and ovary. This is the structure that receives the pollen, rejects or admits the pollen, and eventually produces the seeds in a fruit. Only flowering plants have carpels, and it is undeniably one of the primary reasons why this group contains so many species.

The importance of pollination in generating species diversity

POLLINATOR OR NOT
Not all flower visitors are pollinators; this Australian carnivorous dragonfly is likely hunting pollinating insects.

The discussion here has yet to mention either flowers (aside from the carpel) or pollination. It does appear that there is a correlation between the diversity of flowering plants and the number of different ways a fully functioning flower can be built. Almost every part of the flower can exist in one of a range of character states, although we have to be very careful here not to fall into a circular argument. If we were to classify plants on the basis of floral characters – and they are certainly important – then we cannot take that to mean that the diversity of flowers is responsible for the evolution of that diversity.

The relationship between the number of flowering plant species and the pollination of those flowering plants by the vectors available is exceptionally difficult to unravel. This is because it requires a great deal of lengthy and very detailed fieldwork. This fieldwork must start with the arrival of the pollen on a vector, and it cannot finish until the success of that pollen has been measured. This is more difficult than it might appear for several reasons. First, flower visitors outnumber flower pollinators. Second, if two vectors successfully deposit pollen from a conspecific plant onto the stigma, it cannot be certain that the pollen grains have an equal chance of making it down the style. Third, the seeds have to be harvested and germinated to be certain that the pollination has been successful. This is a lot of research for one population of a species, but the species would have to be investigated across its whole geographical range. Then there are the additional problems that some plants have dozens of potential pollinators, and that some plants are very inaccessible.

The Grant–Stebbins model

The impact of pollinator activity on speciation and thus on increased diversity has recently become a very controversial area of biology. The prevailing theory through the 1970s and 1980s was created by two of the leading biologists of the era, Verne Grant (1917–2007) and G. Ledyard Stebbins (1906–2000). Much later, in 2006, this was labelled the Grant–Stebbins model by Steven Johnson. Since the 1990s, a number of biologists – including Jeff Ollerton, Nikolas Waser, Lars Chittka, Timotheüs van der Niet, Florian Schiestl, Charles Fenster, James Thomson and Scott Armbruster, among many others – have been reassessing the model. Like many biological models, it does work – but only some of the time for the very simple principle that biology is more complicated than we can imagine.

The Grant–Stebbins model for pollination-driven ecological speciation and the overarching work of Stefan Vogel predict that pollinator behaviour drives speciation and species diversity. If this is true, then those orders, families or genera that contain larger than expected numbers of species would be pollinated by a large number of different vectors. The data, however, are inconclusive. Showing that species richness is related to pollinator diversity is difficult for a number of different reasons. For example, a switch back to a former pollinator could have the same effect as taking up with a new species. Another complication is the fact that most flowers are pollinated by more than one pollinator.

Insect pollination has probably been only a secondary (minor) factor in driving speciation, and probably only when the relationship approaches one to one – in other words, when it is very specialised. Having said that, there are species-rich groups of plants that do not have specialised pollinators. An example of the complexity of this problem is found in the work led by Jeff Ollerton into the pollination of species in the dogbane family (the Apocynaceae, which now includes species formerly in the milkweed family, the Asclepiadaceae) published in 2019. What Ollerton has shown is that it is very difficult to see a pattern linking the number of species pollinating a flower, the structure of that flower and the number of species in the genus.

MILKWEED MYSTERIES

Pollination in the plant family Apocynaceae, as represented by swamp milkweed (*Asclepias incarnata*) and the great golden digger wasp (*Sphex ichneumoneus*), is complex but seemingly linked in part to the evolutionary relationships among plants.

Supporting evidence

In support of the Grant-Stebbins model is the fact that most of the 360,000 angiosperm species are pollinated by insects, while the majority of the 1,000 gymnosperms species are wind pollinated. There is also clear evidence, especially from studies on orchids, that pollinator shifts can be associated with speciation events. For example, in South Africa in 2013, researchers found a number of *Eulophia parviflora* orchids that showed variable flower structure. They described two forms that differed in terms of flower structure, fragrance and when they flowered. In particular, the length of the petal spur varied with the different scents. The researchers showed that this difference led to a shift from beetle to bee pollination, and that it was a sufficient change to separate the different forms and to keep them separate.

SOIL SWITCH
Changes in both soil type and pollinator have influenced speciation in the South African iris *Lapeirousia jacquinii*.

In another 2013 South African study, scientists investigated painted petal irises (*Lapeirousia* spp.). The research involved more plants than the orchid work described above and there was a reliable phylogeny (evolutionary tree) for the genus. This is very important, because the modern way to build an evolutionary tree is to use all the data available, including molecular data such as DNA sequence comparisons and protein structures. This makes the tree more reliable and less subjective. Having built the tree you can then map onto it where the different pollinators are active. In the painted petal iris study, the researchers found that there have been 17 pollinator shifts during the evolution of these species and that the shifts were very closely associated with speciation events. However, when the researchers mapped the habitats of the species onto the tree, they discovered that about half of the speciations were associated with changes in the soil type on which the plants were growing. This strikes a warning bell, reminding us that pollination is just one of the possible reasons for the separation of plants, or the splitting of populations that could lead to speciation.

Another argument in support of the Grant-Stebbins model is that pollination by insects will decrease extinction rates by maintaining the gene flow between and within small, sparsely distributed populations. This is relevant because species diversification increases when the creation of new species is greater than the extinction of existing species.

SPURRED ON
This specimen of the orchid *Eulophia parviflora* has short floral spurs and is pollinated by beetles, while the long-spurred form relies on bees.

The problems with the model

Not surprisingly, there are arguments against the Grant–Stebbins model. For a start, insect pollination arose much earlier in the gymnosperms and is still present in the depauperate Gnetales and cycads. Furthermore, plants in the extinct order Bennettitales were insect pollinated and the strategy did not save them. There are obviously aspects of being a plant that trump pollination when it comes to extinction. It is certainly difficult to say whether the increase in insect and angiosperm diversity is a case of co-diversification, because the hymenopterans – the most species-rich group of insect pollinators – were in existence about 100 million years before flowering plants evolved. Another important point, related to pollinator shifts, is that wind pollination has evolved repeatedly from insect-pollinated ancestors, and some of these wind-pollinated groups – for example, the grasses (Poaceae) – are rich in species.

It seems to be true that early insect-pollinated plants tended to be pollinated by many species – in other words they had a generalist flower structure that was accessible to many pollinators. This might have been a factor supporting diversification during this period of evolution. Later on, during the evolution of flowering plants, it seems that more specialised relationships emerged, and the flowers appear to have restricted pollination to one or a few species. This could also have promoted species diversity.

Any argument about the evolution of flowering plants and their pollinators must take into account the prevailing conditions, and these can be difficult to determine and will have varied from place to place in the past, just as they do today. For example, there is some evidence that specialised relationships between orchids and their pollinators may have driven speciation, but at least one piece of research claims to support the idea that an epiphytic lifestyle is a more important factor in facilitating speciation than specialised pollination mechanisms.

ABUNDANT ORCHIDS
Containing primarily insect-pollinated species, the Orchidaceae is one of the largest plant families.

BLOWING IN THE WIND
The grass family (Poaceae) is extremely rich in species, even though these are wind-pollinated.

A multifaceted answer

From the discussion above it is becoming obvious that there are many reasons why there are so many angiosperm species compared to other plant groups, and that insect diversity is one of the secondary reasons. It is important to remember that generating a new species is a different process from maintaining species diversity. Take, for example, the world's five Mediterranean-type regions: the Mediterranean Basin, central California, central Chile, the Western Cape region of South Africa and southwest Western Australia. All five of these repeatedly feature in the world's top 25 list of biodiversity hotspots, or important species-rich areas. The reasons for this extraordinary diversity rarely include pollinator diversity, but they do include variably poor soils; distinct hot, dry summers and mild, wet winters; regular wildfires; variable topography; and constant unidirectional winds.

The answer to the perennial question of why there are so many flowering plants is therefore multifaceted. Flowering plants are extremely speciose compared with other extant groups, though we cannot be sure that in the past other groups have not been very diverse. They are also morphologically and ecologically diverse compared with other extant groups, but there is no key 'radiation-enhancing' innovation such as pollination at the base of the flowering plant branch on the tree of life. The diversity of species across the evolutionary tree of flowering plants is very uneven, and there are different reasons why some lineages have become more speciose. In summary, the causes of angiosperm hyperdiversity are likely to be varied, and both intrinsic to the plants and extrinsic in their habitats. Geographical, ecological and environmental factors all have an influence, and in turn all of these will be influenced by the evolutionary history of the plants present in that place.

MEDITERRANEAN MIX
Regions with poor soils and winter-dry climates often harbour rich plant diversity.

TOP TOME
A primer for pollination studies.

Pollination syndromes

Anyone studying pollination today is soon directed to two highly influential books that synthesised the prevailing paradigm in which pollination was studied. These are *The Principles of Pollination Ecology* (1971) by Knut Faegri and Leendert van der Pijl and *The Natural History of Pollination* (1996) by Michael Proctor, Peter Yeo and Andrew Lack. Both books contain many named examples of pollinator–flower relationships and case studies, and both support the idea that pollination can be neatly classified into pollination syndromes. At the centre of this paradigm is the simple belief that a flower pollinated by a bird will share some basic traits with other flowers that are pollinated by birds. Likewise, the same will be true of flowers pollinated by bats or bees, or wind and so on, until you have named all the different groups of pollinators.

CONVERGENT CACTUS

Hailing from South America, *Cereus repandus* is a member of the cactus family (Cactaceae) and has spiny, ribbed, succulent stems.

EVOLUTIONARY DOPPELGANGER

Also with spiny, ribbed, succulent stems is *Euphorbia abyssinica*, but it originates in Africa and is a member of the spurge family (Euphorbiaceae).

IS EVOLUTION PREDICTABLE?

Convergent evolution gives the impression (which may be true, but equally may be false) that evolution repeats itself. It is a short step from this to the suggestion that evolution is predictable, but this was very much not the view of evolutionary biologist Stephen Gould. In his 1989 book *Wonderful Life: The Burgess Shale and the nature of history*, he argues that if any single major event in the past had not happened – for example, the asteroid impact that precipitated the last mass extinction 66 million years ago at the Cretaceous–Paleogene boundary – then evolution would have proceeded down a different route and he would never have written his book. This is similar to saying that the outcome of a football game would be totally different if the first shot on goal goes in rather than sailing over the bar. While there have been counterarguments to Gould's theory, what cannot be denied is that evolution appears to repeat itself and that sometimes it comes close to being predictable.

The concept of pollination syndromes is attractive to biologists because one of the problems with biology – unlike physics and chemistry – is that it is rarely reliably predictable. Evolution through natural selection may work most of the time, but there is too much stochasticity in living systems for the predictions to be much more than 'probable'. Many biologists are not bothered by this uncertainly and actually enjoy it, while others still want certainty and predictability.

Following on from other ideas contained in Darwin's *On the Origin of Species* was the belief that if two organisms or populations were exposed to the same selection pressures, such as the morphology or behaviour of a pollinator, then the flowers in those two populations would evolve and in time would look similar. This is known as convergent evolution and is seen both in pollination biology and elsewhere in nature. When the similarity is more than just superficial and goes all the way down to the molecular level, it can be referred to as parallel evolution. The key here, however, is that in all examples of convergent and parallel evolution the similarities were arrived at independently.

Types of pollination syndrome

The section below considers the influence of the various pollination vectors on pollination syndromes in an order similar to that in which the vectors were discussed in Chapters 2 and 3. The comments about pollination syndromes are restricted to flowering plants because the vast majority of gymnosperms are wind pollinated, and those that are not (cycads and Gnetales) have already been described in sufficient detail (see Chapter 2).

Abiotic

Water
Where water is the pollination vector, the flowers will be without any specific colour – if they are coloured at all – and there will be no decorative perianth and no scent. The pollen will be released underwater or on the water's surface, but it will not be exposed to the air. The stigma will probably be much divided to increase the chance of catching a passing pollen grain, and the pollen may cling together in the hope of hitting a suitable stigma.

Wind
Wind pollination often involves large quantities of free pollen being released in clouds, or at least small puffs. The presence of a dangling catkin of male flowers is normally a sign of a wind-pollinated plant. The catkins are often accompanied by fewer small female flowers on the same plant, making the species monoecious. When the male and female flowers are on separate plants, making the species dioecious, there are many more female flowers, often in catkins. The individual flowers are normally small, with no coloured perianth and no scent. The catkin-bearing plants tend to be woody.

TALL TREES
Many trees take advantage of their height to shed pollen from catkins, as in common hazel (*Corylus avellana*).

WIND POLLINATION
With their colourless, unscented flowers placed above the foliage, sedges (*Carex* spp.) exhibit the traits typical of wind-pollinated plants.

There are many wind-pollinated herbs, predominant among which are the grasses, followed by the sedges (Cyperaceae) and rushes (Juncaceae). The flowers of wind-pollinated herbs are nearly all bisexual. Maize is a familiar exception and is also a maverick in that its pollen grains have the aerodynamics of a house brick – perhaps to ensure that at least some of the female flowers are pollinated by the male flowers carried above them. The flowers of wind-pollinated herbs tend to be dowdy, with much-reduced perianths and no scent.

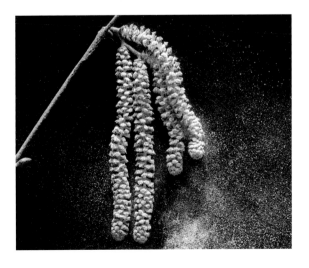

Different but similar

A simple but powerful example of the difference between wind pollination and insect pollination involves the plantains (*Plantago* spp.) and foxgloves (*Digitalis* spp.). Until the end of the twentieth century, the former were placed in the family Plantaginaceae and the latter in Scrophulariaceae.

At the time, Scrophulariaceae was a catch-all family of closely related genera but was probably not a monophyletic group, which is now the only type of group allowed in plant taxonomy. With the addition of molecular data and microscopic details, the families on this branch of the evolutionary tree have since been sorted out (almost). And it turns out that foxgloves and plantains are, in fact, closely related and therefore should be in the same family.

This seems unlikely at first glance, because the plantains are drab, with a tiny brown perianth and much-divided stigma, while the foxgloves have large colourful flowers with a small two-lobed stigma. However, if you count the number of parts in the four whorls of the flowers and record how they are attached to one another, the floral formula (as it is known) is the same for both.

YELLOW FOXGLOVE
Digitalis lutea.

RIBWORT PLANTAIN
Plantago lanceolata.

MIXED MESSAGES
French meadow-rue (*Thalictrum aquilegiifolium*) flowers are pollinated by both wind and insects, so don't conform to a specific syndrome.

where a wind-pollinated species appears within an animal-pollinated group. Sometimes we find species that may be going through this process in front of our eyes, as in the beautiful meadow-rues (*Thalictrum* spp.) in the buttercup family (Ranunculaceae). Despite conforming to some of the expectations of insect pollination, including being very colourful, some thalictrums are pollinated by both wind and animals.

Vertebrates

Lizards

Lizard pollination is rare and often associated with oceanic islands, where the reptiles can fill ecological niches that in continental areas are filled by small mammals. It is very difficult to see any defining set of traits in plants that are pollinated by lizards, likely because the reptiles are generally omnivores and the food provided by plants is just a small part of their diet. The flowers do have to be reasonably robust, however, and preferably not attractive as a meal, because the larger the pollinator, the greater the risk that the entire flower will be eaten.

Wind is blind, so signals aimed at attracting it as a pollinator are wasted. And as it is largely uncontrollable (except close to the female parts – see page 58), it needs no reward or incentive to visit. Wind pollination is therefore a straightforward syndrome – although there are exceptions to the rule, including the many examples

DISTASTEFUL DEFENCE

One of the best-documented examples of lizard pollination is that of the tree spurge (*Euphorbia dendroides*) on Crete. Like other members of the genus, it has very stripped down flowers with no perianth, but with conspicuous anthers, stigma and nectaries that are located close to each other in a pseudo-flower. The flowerheads do not apparently restrict access to any animal, but there is a very simple reason why they are not eaten wholesale: all spurges produce a caustic white latex. A few animal species have evolved a mechanism for avoiding the toxicity, but most – including goats – leave them alone after experiencing the burning sensation in their mouth. It seems that the lizards are tougher than most, and that the latex does not bother them.

LIZARD POLLINATION
Few features typify lizard-pollinated flowers, but they must be robust, as in tree spurge (*Euphorbia dendroides*).

TYPES OF POLLINATION SYNDROME   169

Non-flying mammals

When compared with lizard pollination, non-flying mammal pollination is well studied. This is due entirely to biologists' obsession with mammals and nothing to do with their global significance in pollination, nor the ease with which they can be studied – many of them are nocturnal and dwell in dense vegetation. Like lizard pollinators, non-flying mammalian pollinators derive much of their nutrition from non-flower sources, and so there is not the same level of co-dependency as found in other pollination syndromes. One of the problems for a non-flying mammal that is eating pollen or drinking nectar is that it is a sitting target for predators. Furthermore, many of the flowers that are pollinated by these small mammals are held at ground level, making the animals even more vulnerable. This is particularly evident in Australia, where ground-dwelling possums are important in the pollination of some of the banksias (*Banksia* spp.). The vulnerability of the possums has come to light following the introduction of cats, rats and foxes to Australia, with devastating repercussions.

One of the reasons for the paucity of pollination by small mammals may be that they are unable either to defend themselves or to escape fast enough. However, at night they appear to be safer and so this is when they come out, presumably attracted by scent and maybe white flowers on moonlit nights. The reward has to be commensurate with the size of the animal, and so flowers pollinated by small non-flying mammals can look very similar to some of the flowers pollinated by birds. The mammals do not compete directly with the birds, however, because they are nocturnal and the birds are diurnal. And they do not compete with flying mammals – the bats – because these animals almost always visit flowers that are held high in the air.

Bats

If you were looking for a group or clade of pollinators to support the concept of pollination syndromes, you might well be tempted to start with bats. At first sight, flowers pollinated by bats appear to conform to a pattern. For a start, they are robust, because delicate little flowers would be damaged by a relatively clumsy bat. This seems like a harsh way to describe such an agile flyer, but once they have landed, bats do have to hang on tightly. The flowers also tend to be very exposed, either on a tall stalk or protruding beyond the leaves of the tree on which they grow. They will generally smell fruity, bordering on fermented, and there is likely to be plenty of pollen and very large volumes of nectar, of lowish concentration and generally unprotected by a long petal tube. The flower colour is likely to be dull greenish to

NIGHT SHIFT

Scarlet banksia (*Banksia coccinea*) flowers welcome bird pollinators during the day and mammals like this western pygmy possum (*Cercartetus concinnus*) at night, doubling the chance for pollination.

yellowish, and the individual flowers sometimes last just one night. Some plants even have a saucer-shaped leaf or petal, modified to aid with echolocation.

Just when it starts to look like Darwin was right and that pollination syndromes exist and are the rule, the bat-pollinated jade vine (*Strongylodon macrobotrys*) comes along. The flowers of this member of the bean family (Fabaceae) are generally pollinated by bees. Typically, the insects have to push into the flower for the nectar, and in the process force the stamens out of hiding and are dabbed with pollen. However, the jade vine produces very large volumes of nectar that can be seen glistening in the bowl created by the sepals, and the colour of the flower is an artificial-looking eau de Nil colour. It could be that the bats think that this colour is dull but they are attracted to it and get their reward. Perhaps significantly, a similarly artificial but darker colour is produced by *Puya* species, although sadly the study of pollination in this genus is uneven and incomplete.

ON THE BILL

The New Zealand native kākābeak (*Clianthus puniceus*) gets its name from the kākā (*Nestor meridionalis*), a parrot whose beak shape resembles the curving flowers. While kākā do visit the flowers, it is likely that native honeyeaters – tūī (*Prosthemadera novaeseelandiae*) and bellbird (*Anthornis melanura*) pollinate them, yet the rarity of the plants may hinder the ability of the birds to enact cross pollination.

A bird-pollinated flower from New Zealand further dispels the theory that pollination syndromes are anything more than just recommendations rather than guidelines. The flowers of kākābeak (*Clianthus puniceus*) are a dead ringer for those of the jade vine, except they are cerise and the plant is pollinated by birds (see box on page 42). When you look at the basic structure of bat-pollinated flowers, they can be anything from the open bowl of an agave (*Agave* spp.) to the strongly bilateral flower of the jade vine, so the original assumption that bat-pollinated flowers would conform to a simple syndrome was incorrect.

CROSS PURPOSES

Pollinated by bats and bees, the garish blooms of jade vine (*Strongylodon macrobotrys*) conform to neither syndrome.

Birds

There does not appear to be a clear syndrome for the flowers of bird-pollinated plants. As already shown, they have some features in common with flowers that are pollinated by mammals, both flying and non-volant. And as birds are a very diverse group of 10,000 species, it might be over-optimistic to think that the flowers they pollinate will be similar.

Birds generally do their pollinating during the hours of daylight – they are diurnal. The flowers are generally brightly coloured, as are the birds themselves. Red is often assumed to be the default colour for bird-pollinated flowers, but this is not strictly true. However, if you add orange to that you might be describing a majority of bird-pollinated flowers. The question of scent is controversial, with ornithologists disagreeing on the olfactory sensitivity of birds. If bird-pollinated flowers do smell, then it is only a weak smell compared with many bat-pollinated flowers. Similarly, bird-pollinated flowers contain less pollen than bat-pollinated flowers and the nectar is more likely to be at a weaker concentration.

The flowers themselves often have tubes made up of fused petals or other parts, but this is thought by some biologists to have as much to do with excluding nectar thieves as with the selection of birds. This raises the knotty question of whether the relationship between the birds and the flowers they pollinate really is a mutualism. There is a good argument to be made that pollination is, at its heart, simple herbivory (see page 82).

Invertebrates

Moths

Moths have a lot to answer for in the debate about the significance of pollination syndromes. Possibly the biggest red herring in biology is Morgan's sphinx moth (*Xanthopan morganii*), a Madagascan species with a ludicrously long proboscis that enables it to harvest the nectar at the base of the spur that protrudes from the labellum of Darwin's orchid (*Angraecum sesquipedale*). Because Charles Darwin accurately prophesised the existence of this moth based on the combination of the flower's spur and its nectary, the white perianth and the night scent, biologists were happy to apply this to all 360,000 species of flowering plants. With hindsight, this was an over-optimistic conflation.

That said, moth-pollinated flowers are often scented at night – if they are pollinated by nocturnal species. However, not all moths are nocturnal, including the burnet moths (Zygaenidae), the hummingbird hawkmoth (*Macroglossum stellatarum*) and the cinnabar moth (*Tyria jacobaeae*). The scents of moth-pollinated flowers can be very strong if they are pollinated at night – as in

> **RED HOT**
> Bird-pollinated flowers are often red or orange, as here in the Christ's tears tree (*Erythrina crista-galli*), because birds are typically day-flying and need a visual cue.

NIGHT FLIGHT
Honeysuckle (*Lonicera periclymenum*) attracts moths with scent.

the case of honeysuckle (*Lonicera periclymenum*) – but day-active moths may be attracted by flower colours. The flowers visited by nocturnal moths tend to be pure white or white with pale tints of yellow or green. There is normally a tube, which is longer in flowers serviced by hummingbird hawkmoths. The nectar varies in strength and volume, with hummingbird hawkmoths needing larger volumes of more concentrated nectar.

Butterflies

Butterflies are diurnal, and so not surprisingly the flowers they visit differ from those visited by nocturnal moths but are similar to those pollinated by diurnal moths. Like birds, butterflies generally gravitate towards warm colours, such as yellow, orange and red, and they also prefer to be rewarded with weakly concentrated nectar. Being smaller, butterflies require less nectar overall than birds, but proportionately the quantity is the same for both. Finally, butterflies respond well to smells that are pleasant to us. It is very clear from this that some flowers – including common garden species such as purpletop vervain (*Verbena bonariensis*) – will attract a wide range of butterflies.

LANDING PLATFORM
Flat-topped flower clusters provide a ready landing pad for butterflies.

Beetles (excluding carrion beetles)

Some beetles are attracted to the smell of dung, faeces and rotting carcasses, and these are considered below. The rest of the beetles that are involved in pollination are attracted to sweet and spicy scents, and seem to be quite indifferent to flower colour. There is less variation in the shape of beetle-pollinated flowers than in most other groups. The flowers are mostly dish- or bowl-shaped, with the nectar exposed, which means that the nectar can be removed by visitors that do not effect pollination. However, for many beetles the reward is not just the nectar, nor the chance to meet a mate, nor the extra warmth (see Chapter 5), but the pollen. As the insects are quite messy eaters, they get coated in the pollen and then take it to the next flower. Therefore, it is common for beetle-pollinated flowers to produce seemingly excessive quantities of pollen, second only to bat-pollinated flowers.

Carrion flies and beetles

These two groups are often treated together, but this may be as flawed as considering all bees together, because flies seem to be more easy duped and satisfied than beetles. Putting this more clearly, carrion flies will visit (and pollinate) a flower simply in order to lay their eggs on the malodourous flower parts. The unfortunate maggots discover too late that the flower has fooled their inattentive parents.

Beetles, on the other hand, visit foul-smelling inflorescences like those in the arum family (Araceae) not to lay their eggs but to find a mate. The extraordinary relationship between lords-and-ladies (*Arum maculatum*) and its pollinators (see page 39) is, however, the exception, and many relationships between flowers and their beetle pollinators are as run of the mill as any plant–herbivore relationship. This was ignored for many years, however, because kicking this proverbial hornets' nest seemed to be anti-Darwinist. In fact, the study of pollination has shed more supporting light on evolutionary principles and processes than it has contradicted Darwin's proposals (see page 180).

Flies

If there was an award for the group of pollinators that is most commonly acknowledged and then dismissed, it would be handed to the flies. Like the beetles, the flies are another large group, but one that rarely forms specialised relationships – unless they are the flies that are attracted to the smell of dung, faeces and dead smelly carcasses (see above). Flowers that are pollinated by flies tend to be rather dull, like the flies themselves, being white or in light shades of already pale colours. Scents are either absent, or faintly musty or fungus-like. Flies do not seem to be very inquisitive in this instance, and choose flowers where the pollen or nectar is easily accessed – normally ones that are bowl-shaped. The flat-topped inflorescences of members of the carrot family (Apiaceae) are often used to illustrate the fly pollination syndrome, but there are many fly-pollinated flowers in other families, and other groups of insects pollinate flowers and inflorescences conforming to the fly-pollination syndrome. There is a good deal of similarity between beetle- and fly-pollinated flowers, and the only major difference is that beetles tend to follow stronger scents.

POLLEN FEAST
Montpelier rockrose (*Cistus monspeliensis*) flowers use abundant pollen to attract beetles.

Bees

Bees, the poster animals of pollination, are perhaps the most difficult to associate with a particular flower structure because there are so many of them and there is an extraordinary diversity within the approximately 20,000 named species. In fact, grouping all bees together is perhaps as illogical as grouping all 10,000 birds together. The very specialist scent-collecting euglossine bees were discussed in Chapter 5, and these are quite distinct from the more familiar honeybees.

The many books and journal papers on pollination do not present a united view of what colours bees prefer. One will say that that the insects prefer 'lively' colours , especially blues and yellows, while another will say that they also pollinate white, pink and purple flowers. It seems that only red and the closely related orange are not favoured by bees, but could this be because they are rarer colours – or are they rarer because bees ignore them?

Some of the literature states that bee-pollinated flowers often require the bee to deliberately enter some type of chamber in order to access the nectar. This is certainly true for some irises (*Iris* spp.), snapdragons (*Antirrhinum* spp.) and legumes (Fabaceae), but there are many other floral architectures that bees visit, making it very difficult to discern a particular shape that is favoured by the insects. The same is true for scent and reward. The only floral feature that is most likely to be associated with bee pollination is some type of nectar guide, which may be either visual or olfactory.

BEE DIVERSITY
With so many different types of bees, the flowers they pollinate are equally diverse.

Ants

It has already been suggested that ants are not well suited to being pollinators for a variety of reasons (see pages 92–3). As a result, few pollination scientists even attempt to describe a pollination syndrome for ants. However, there is a combination of characters that, when found together, does demand that ants are considered. Small, scentless green flowers held near the ground, or on easily climbed plants, can be visited by ants and hence pollinated by them. The nectar of these flowers has to be easily accessible, because ants do not like to hang around in places that leave them exposed to predators.

Wasps

Flower-visiting wasps are ignored even more than ants, probably because the majority of them are just visitors rather than pollinators. However, if you sit quietly in your garden you may well see wasps visiting a similar group of flowers to those visited by the bees.

WILY WASPS
While wasps are frequent flower visitors, little is known of their effectiveness as pollinators. This is a hornet (*Vespa crabro*), a large species of wasp.

Do pollination syndromes still exist?

In 1996, a scientific paper authored by Nickolas Waser and colleagues proposed that the assumption that evolution favours a trend from generalist to specialist pollination relationships should be reassessed. The authors also questioned the cherished belief that pollination syndromes prove that specialisation is the inevitable journey's end for evolution. This theory is similar to the law of competitive exclusion, which states that two organisms cannot occupy the same ecological niche without one of them triumphing. This may be true in a contrived laboratory experiment, but it is not true in the field. If it were, then tree branches in the tropics would not be providing homes to dozens of different epiphytes. The Waser group pointed out that there are many examples of flowers that are pollinated successfully by many species, and many species that successfully pollinate many flowers. Furthermore, they noted that the ranges of plants and their pollinators do not always overlap, so one plant species may have different lists of pollinators in different parts of its range.

Since the publication of that groundbreaking paper, there has been a lively debate on the relative importance and occurrence of specialised pollinators, specialised flowers, generalist pollinators and generalist flowers. It has been suggested more than once that the language and choice of words used has been confusing. However, some clarity does seem to be emerging. Papers led by Scott Armbruster (2000 and 2017), Charles Fenster (2004), Jeff Ollerton (2007 and 2009) and Víctor Rosas-Guerrero (2014) have batted ideas back and forth. In all probability, everyone is correct in part.

Biologists can now say with certainty that pollination relationships are changing. In 2012, Timotheüs van der Niet and Steven Johnson reviewed the ways in which plants could change their pollinators; their results are shown in the table below. The table shows that there have been many pollinator shifts. But what it is does not show is how many animal-pollinated lineages have become wind pollinated, or both wind and animal pollinated. From what we know to date, it appears that the switch from animal to wind pollination is not uncommon, but that a switch in the opposite direction is rare. For example, researchers have shown that two species within the predominantly wind-pollinated sedges have switched to animal pollination. What the table shows is that being a generalist should be considered a syndrome.

POLLINATOR SHIFTS

Yellow boxes indicate that there has been a shift from the pollinator on the left of the row to the pollinator at the top of the column; pale yellow boxes indicate a pollinator shift within the guild of pollinators – for example, from one wasp species to another species of wasp (from van der Niet & Johnson 2012).

from \ to	Selfing	Moth	Day hawkmoth	Wasp	Bat	Fly	Bee	Bird	Non-flying mammal	Long-proboscid fly	Generalist	Butterfly
Selfing												
Moth		■			■	■		■	■	■	■	■
Day hawkmoth	■											
Wasp	■			■								
Bat			■				■				■	
Fly												
Bee							■					
Bird			■			■		■				
Non-flying mammal											■	
Long-proboscid fly											■	
Generalist					■		■	■		■		
Butterfly	■						■					

It is clear that specialisation comes in a number of different forms. As summarised by Scott Armbruster in 2012, specialised pollination is where a plant is pollinated by one to a few animals in the same group, ecological specialisation is where a pollinator visits one to a few flowers for food, floral specialisation is where there are features of the flower that limit or restrict visitation, and evolutionary specialism is the change from a generalist to a specialist, which according to the table above has happened four times (as indicated by the yellow cells).

Likewise, it is clear that there are many different ways to be a generalist flower, as described by Jeff Ollerton and colleagues in 2007. Some generalist flowers provide a reward to many visitors that are all able to pollinate the flower, whereas others have easily accessible rewards but are visited by only a small number of pollinators. Some generalists attract many visitors but only a few of these pollinate the flower, while some flowers look as if they are specialists but in fact are visited successfully by many pollinators. Finally, and as mentioned above, there is a group of plant species that are pollinated by different animals across their range.

Summing up the syndromes debate

So do pollination syndromes exist? The answer is yes, but not every plant or pollinator has signed the pledge. The best way to illustrate how difficult it can be to categorise a plant and its flowers is to use an example.

A team led by Steven Johnson, one of the world's current leading pollination biologists, carried out work on aloes (*Aloe* spp.) in South Africa in 2006. There are more than 500 aloe species in Africa and neighbouring areas, and many are pollinated by birds. This is not surprising, because they have tubular flowers with copious nectar. Normally this nectar is clear, but in a small number of species, including *Aloe vryheidensis*, the nectar is dark brown due to the presence of phenolic compounds, which also give it a bitter taste. The bitterness deters some of the classic long-billed birds that pollinate flowers in this part of the world, such as sunbirds (Nectariniidae). However, it does not deter a range of birds – including bulbuls (Pycnontidae), white-eyes (Zosteropidae), rock thrushes (*Monticola*

COLOUR BLIND
Dark-capped bulbuls (*Pycnonotus tricolor*) readily consume the brown nectar of some aloes.

spp.) and chats (Saxicolinae) – that also feed on insects and fruit, and for whom the nectar is part of a mixed diet. The obvious question then is why are these species not put off by the phenols? The answer may be that they already tolerate these flavours in the insects and fruits that they eat – in other words, they have an exaptation to bitter-tasting food.

The bitter nectar is unpalatable to bees, but when birds were excluded by Johnson and his team from *Aloe vryheidensis* flowers, the bees were still able to access the pollen and while removing this they would successfully pollinate the flowers, all without going anywhere near the bitter nectar. What is seen here is a close relationship – a mutualism – for part of the year between *A. vryheidensis* and several bird species with short bills that are not run-of-the mill bird pollinators, and that also eat insects and fruit. The same flowers that are pollinated by these birds are also pollinated by bees. The dark-coloured nectar is an honest signal that deters nectar thieves, which include the bees that actually do pollinate the flowers but only when they are taking pollen. This example illustrates that one flower may exhibit two pollination syndromes simultaneously, thereby drawing into question the reliability of predictions based on the syndrome concept.

SIDE-BY-SIDE SYNDROMES
Aloe vryheidensis attracts bees with pollen and birds with nectar.

On the other hand, research carried out in 2002 by Peter Goldblatt and John Manning on another predominantly African genus, *Gladiolus*, supports the concept very well. The gladiolus species that are pollinated by moths are relatively large (for the genus), with a distinctly long tube, pale coloration (including a dull purple brown) and strong scent. The flowers open fully only at night-time, and their nectar is rich in sucrose and highly concentrated. The gladiolus species that are pollinated by butterflies are also relatively large (for the genus) and have a distinctly long tube, but they are red with white markings. Their nectar is less concentrated than that of the moth-pollinated flowers and it is rich in sucrose or glucose/fructose. The evolutionary tree for *Gladiolus* appears to indicate that moth pollination has emerged independently six times and butterfly pollination three times. This is a good example of convergent evolution, but from what did the plants evolve?

The two similar pollination syndromes in *Gladiolus* species have not simply been switching between each other. Instead, it seems that the moth-pollinated species have evolved from bee-pollinated ancestors, while the butterfly pollinations arose in species that were pollinated by either bees or birds. This implies that it is not difficult for plants to shift from one pollinator to another, and in the whole of *Gladiolus* there have been 32 pollinator shifts. The existence of pollinator shifts means that the lines between different syndromes is inevitably going to be blurred. It also means that, as the climate changes, some plants and animals may be better suited to tolerate these changes.

SYNDROME SWAP

Gladiolus liliaceus is one of the South African species studied in Peter Goldblatt and John Manning's seminal paper published in 2002.

Aspects of evolution illustrated by pollination

As pollination biologists examined more and more species, it became clear that convergent evolution is at work here. The fact that pollinator type can sometimes be successfully guessed from the structure of the flowers of a taxonomically diverse group of plants shows that it happens. For example, tubular red flowers pollinated by birds are seen in many plant groups, and thermogenesis is linked to providing a mating arena in beetle-pollinated species.

A very nice example of convergent evolution is seen in the protea family (Proteaceae). Plants in this group in Australia and South Africa have independently switched from being pollinated by birds to being pollinated by ground-dwelling mammals, but species that are very different. In Australia, the pollinators are marsupials, whereas in South Africa they are rodents. In both cases the flowers are produced at ground level.

Advergent evolution is another evolutionary event in which two very different organisms share a common trait. It differs from convergent evolution in that one species does not change but the other one does, thereby becoming increasingly similar to the first. The classic example of this is the bee orchid (*Ophrys apifera*), whose labellum looks increasingly like a female bee but the female bee does not change in any way. This can also be called mimicry, but advergent evolution is more accurate in this context. Another example can be seen in the broad-lip bird orchid (*Chiloglottis trapeziformis*) and the thymine wasp *Neozeleboria cryptoides*. Both produce the molecule 2-ethyl-5-propylcyclohexan-1,3-dione, which is known nowhere else in nature. This is not convergent evolution because the female wasps were producing this sex pheromone to attract male wasps long before the orchid started to produce the same compound.

The speed of evolution is another part of the story Darwin wrote about, and he was sure small changes were always acquired slowly. This does not appear to be the case, however, and it seems that evolution can speed along – which is lucky if the climate is changing quickly. Those changes might include switching from one pollinator to a new one, or even shifting from one pollination syndrome to another.

As mentioned throughout the chapter, there has been a renaissance in the study of pollination in the past 25 years. This has been helped by technological innovations such as high-powered microscopy and the ability to sequence the DNA of a species overnight, which has allowed biologists to build increasingly accurate evolutionary trees, or phylogenies. These have revolutionised the study of evolution. For example, dates can now be put on the emergence of characters in both plants and pollinators. It is also possible to see whether there has been co-radiation within species-rich groups, especially those with specialised mutualistic relationships.

Pollination is an excellent field of biology to study as part of an investigation into the progression of evolution.

BIRD BANQUET
The South African king protea (*Protea cynaroides*) is bird-pollinated.

CHANGE OF DIRECTION
The bird-pollinated scarlet banksia (*Banksia coccinea*) in Australia also attracts mammals.

Chapter summary

As a result of work in this area, we now have a much better-informed answer to some big questions, including why there are so many species of flowering plants.

The concept of pollination syndromes has become highly controversial. Pollination relationships include good examples of convergent evolution and advergent evolution, and many people have argued about the importance of specialisation in these relationships. However, just like life itself, the answers are not clear cut and may be different in different times and different places.

Chapter 7

Pollination, Pollinators and People

Pollination has a key position in the life histories of the majority of plants on Earth today. Of these, somewhere in the region of 85–90 per cent of flowering plants are pollinated by animals. This means that without pollinators the ecology of Earth would fall apart. This could be the shortest chapter in this book based on these three sentences. However, if we are to appreciate the need to look after pollinators, we need more detailed information on the products and results of pollination.

Sweet almond (*Prunus dulcis*).

PREHISTORIC POLLINATOR?
Preserved in amber dating from the early Eocene (44.1–47 million years ago), *Palaeomyopa* sp. is a fly in the family Conopidae; some modern conopid flies parasitize bees, wasps and ants.

Measuring the value of pollination

The ways in which pollinators support our lives are many and varied, and these can be grouped together in different ways that sometimes overlap. There can be inherent tensions between the different ways of cataloguing and discussing nature, and the same holds true when it comes to investigating the value of pollination.

In the past two decades there has been a big effort to place a price tag on the different components of nature. However, putting the future of the Earth in the hands of economists is anathema to some people, as the monetarisation of nature is diametrically opposite to the assumption that it has an intrinsic value that is beyond price. Critics of this latter view have tried to belittle the 'tree-huggers' on the grounds that we live in a secular, neo-liberal world where utopian ideals have been found wanting. Perhaps surprisingly, the neo-liberals follow the words of the Book of Genesis, which state that the Earth is here for the use of mankind and that all other species are here at the disposal of mankind. Disposal is something that mankind is good at right now, and we need to appreciate the potential effects of carelessly depleting the world stocks of pollinating animals.

A critical dependence

One approach to exploring the effects of pollination is to consider what Earth would look like without it. Evidence of pollination is first seen in the fossil record about 170 million years ago, so any plants that have evolved since then – including all the flowering plants – would not exist. However, the evolution of gymnosperms as we know them today has also been determined by wind and animal pollination, and so they would not exist either – and nor would humans, according to those people who support the work of evolutionary biologist Stephen Gould (see box on page 166). Whether this is true or not, there is no denying that animal pollination is important – or even 'critically important', as stated by biodiversity scientist Jeff Ollerton. This chapter looks at some of the ways in which our existence is dependent on pollination and thus pollinators.

VALUING NATURE
Pollinators are essential for our continued existence. Without them, many crops would fail to fruit, a point of importance when putting a price on nature.

Words are important

This might seem like and a strange and unnecessary statement 54,000 words into a book, but this chapter is looking at pollinators in terms of their relationship with humans. The relationship between a pollinator and a flower is sometimes tricky to name. Is it a mutualism or is it closer to parasitism? Is the nectar a reward or inducement, or is it payment for a job well done? When the relationship with humans is discussed, words like function, service, benefit, worth and value are used. Value is particularly tricky, because a valuation often has to include a price, and price implies that the object is owned and potentially up for sale. In nature, this is clearly nonsense. The sun is very valuable to life on Earth, because without it, photosynthesis would not have been an option for fuelling this life. Yet no one would try to put a price tag on the sun.

organisms, provide services that are beneficial and valuable to humans.

The same paper gives a classification of ecosystem services, showing where they fall in the cascade. These services are split into three groups: provisioning services, including food and other resources required by animals, plants, fungi and other organisms; regulating and maintenance services, such as geochemical cycles and erosion control; and cultural services, which include all the uniquely human services that do not fall in the previous two categories. These groups are not mutually exclusive – for example, pollination and fruit dispersal are found under regulating and maintenance services, but clearly pollinators are also very important components of provisioning and cultural services. As stated earlier in the book, the words of physicist Erwin Schrödinger frequently come to mind when studying pollination: not only is it more complicated than we imagine, it is also more complex than we can imagine.

Philosophy, biology and economics

A simple way to group the ways in which pollinators impact on our lives is in the three Es: economic, ecological and ethical (plus aesthetic). Other biologists have used the three Ps – practical, pragmatic and philosophical – while a third group employs the three Ss – selfish, sensible and spiritual. A more academic way to look at the same divisions is practical, theoretical and abstract.

Another framework for analysing and dissecting the relationship between nature and humans was proposed in 2017 by Alessandra La Notte and her colleagues. In this cascading framework the biophysical structure of nature includes components with functions that work together in many interactions and in different ways. These functioning components, including the

SERVICE ECONOMY
The intricate relationship between pollinators and plants is subject to valuation as an ecosystem service.

IS THIS REALLY A MATTER FOR THE UNITED NATIONS?

Anyone doubting the importance of pollination should look at the report *Pollination, Pollinators and Food Production*, produced in 2017 by the Intergovernmental Science-Policy Platform on Biodiversity and Ecosystem Services (IPBES). This 556-page document is a very thorough assessment of the role of pollinators, the problems facing them and the ways in which they can be supported. The need for such a report was identified in 2002 at the sixth meeting of the countries that have signed up to the aims of the 1992 Convention on Biological Diversity (essentially everyone apart from the US).

The IPBES report states that the role of pollinators is so important that the United Nations has said that 'pollination is one of the most important mechanisms in the maintenance and promotion of biodiversity and life on Earth'. This is very powerful stuff, and as a direct result of the IPBES report the world's governments (except the US) signed up to the International Pollinator Initiative, which will run until 2030. This is a bold attempt to protect and restore pollinators and is looked at in more detail in Chapter 8.

IT'S OFFICIAL
The United Nations report on the value of pollination services.

The aesthetic value of pollination

The value of pollination in terms that are not primarily linked to economics or ecology are often tagged on at the end of any account of the subject. This instantly devalues them, so in this book the aesthetic, cultural and spiritual values of pollination are looked at first.

The evolution of pollination has enhanced the quality of our lives in many areas, and high up on this is gardening. Gardens without flowers may be trending at the Chelsea Flower Show, but they will never replace borders of blooming specimens if the public reaction to the garden that won the best in show in 2019 is anything to go by.

The benefits of gardening are many, including physical fitness and metal relaxation. Gardens can be deeply spiritual places. In the opinion of the English poet Dorothy Frances Gurney (1858–1932), 'One is nearer God's heart in a garden | Than anywhere else on Earth'. One is also able to get very near to pollination in a garden, and it is not just the animal-pollinated plants that can give pleasure. The flowers of two common garden grasses, maiden silvergrass (*Miscanthus sinensis*) and giant feather grass (*Stipa gigantea*), can be examined closely, even without a hand lens. Natural selection has driven the construction of a beautifully simple structure to promote wind pollination. There may be neither scent nor colour, but there is great beauty in these flowers.

Then there is the scent of a garden. Who can deny the attraction of the scents of roses? There is a genuine sense of disappointment when a rose has no scent – it is just wrong. In the dead of an English winter, normally on Christmas Day, the flowers of Chinese witch-hazel (*Hamamelis mollis*) emit an aroma that makes Chanel No. 5 smell like an antiseptic. Even the revolting smell arising from the dragon arum (*Dracunculus vulgaris*) has a certain attraction. Pollinators remember scent, and it can also provoke very powerful memories in us. The smell of the flowers of mock orange (*Philadelphus coronarius*) takes me straight back to my childhood 50 years ago. Many floral scents are used in aromatherapy for relaxation and they can perform the same service free in the garden.

WINTER WONDER
The rich floral fragrance of Chinese witch-hazel (*Hamamelis mollis*) enlivens gardens in the depths of winter.

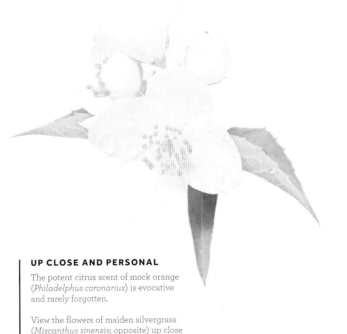

UP CLOSE AND PERSONAL
The potent citrus scent of mock orange (*Philadelphus coronarius*) is evocative and rarely forgotten.

View the flowers of maiden silvergrass (*Miscanthus sinensis*; opposite) up close to see how wind pollination works.

The cultural importance of pollination

The connection between bees and humans is particularly deep. We describe busy people as being 'busy as bees' and we explain reproduction to our children under the euphemism 'the birds and the bees'. The love of humans for bees is proven by the number of people who provide homes for solitary bees in bug hotels and those who practise beekeeping as a hobby. Bees are reputed to be able to smell fear in humans and sense imminent disaster. Whether these skills are reliable is unknown, but the insects certainly intrigue humans with their communication systems and their societies. Scientific careers have been built on understanding the genetic foundations of the eusociality of bee colonies and bee behaviour.

Our lives are much richer thanks to the pollinators around us, and a garden without bees, butterflies and all the other pollinators is lacking an essential dimension; it is somehow less alive. The aesthetic value of pollination certainly benefits people who garden, but it is not restricted to gardeners. One of the most popular areas of London Zoo in Regent's Park is the Butterfly Paradise exhibit, where you can get very close to some beautiful butterflies. However, you may have difficulty seeing some of them at first because of the extraordinary

BUG HOTEL

By providing homes for pollinators like mason bees (*Osmia* spp.), gardeners can improve pollination of their crops. Another benefit is the joy of observing the life cycle of these fascinating creatures.

lengths they go to in order to camouflage themselves. The delight on the faces of visitors, both young and old, should convince anyone of the spiritual and cultural value of pollinators and, in some cases, their extraordinary larvae.

THE VALUE OF TRADITION

It is important to note that untangling the cultural and economic values of pollination can be difficult. Take, for example, holly (*Ilex aquifolium*) and mistletoe (*Viscum album*). They have both been part of winter celebrations since pagan times, and yet neither could honestly be described as an essential part of anyone's life. Even so, they are now iconic Christmas plants, recognised especially for their berries – so much so, in fact, that branches lacking berries are worth only half to two-thirds of those with them. The monetary value may be small, but our lives are richer.

The ecological value of pollination

FIELD FRINGES
Leaving a strip of uncultivated land alongside a field increases plant, and therefore pollinator, diversity.

It could be argued that pollinators are to nature what wheels are to a vehicle: they keep it moving forward. There is, of course, more than one type of tyre that you can put on your car, giving you the opportunity to use that car in more than one situation and resilience in the face of different weather conditions. Different vehicles need different-sized tyres and society currently needs all of those different vehicles to function. In ecological terms, the full range of pollinators are seen as providing maintaining, supporting and regulating roles.

In the list of ecosystem services drawn up in 2017 by Alessandra La Notte and colleagues (see page 185), pollination is described as a service that interacts with other services. Some of these other services are benefits, such as cultivated crops; some are regulating and maintenance services, such as the erosion-prevention services provided by seagrasses and carbon dioxide sequestration; and some are interactions, such as bioremediation, control of water flow and prevention of soil erosion. Wherever you look in nature, pollinators are providing functions that are of great value, benefit and service. Pollinators are also important for the future through the evolution of new varieties and species.

The conservation of pollinators is dealt with in detail in Chapter 8, but it is clear here that if you look after pollinators, other ecosystem services improve. This very nicely and simply shows the central position of pollination within ecology. For example, in northern temperate agricultural landscapes, a 6 m-wide (20 ft) strip around a field is commonly left uncultivated and filled with prolifically flowering native plant species to provide food for birds. Fields are separated by equally prolifically flowering woody plants in hedges that provide nesting and feeding opportunities for birds.

While the funding of such projects may have been to support birds, the habitats created provide many other gains. For example, the birds and other organisms (such as spiders) that thrive here provide important crop pest-control services, which in turn has the further benefit that pesticide use can be reduced. In some circumstances the inclusion of nitrogen-fixing leguminous plants in these strips can be used to improve soil fertility, and legumes are nearly all animal pollinated. The hedges, meanwhile, play a very important role in soil protection. Soil is often overlooked in landscape planning, but erosion control is essential if agricultural production levels are to be maintained. Hedges and the uncultivated strips around fields are also very important in controlling water run-off following heavy rain. So, what starts off as a simple project to support birds or pollinators ends up having myriad ecosystem benefits.

MISTLETOE
Viscum album (left).

The ecological knock-on effects

One of the important factors to consider when trying to identify the ecological role of pollinators is whether the benefits they provide – for example, flood control – might also be experienced where they do not occur. This can be illustrated with the example of two plants that share a specialist pollinator. Brazil nut trees (*Bertholletia excelsa*) and *Stanhopea* orchids are pollinated by the same euglossine bee species. The male bees harvest scent from the orchid, without which it is less likely they will mate. However, it is the females that pollinate the flowers of the Brazil nut tree. This is why Brazil nut tree plantations that do not also support *Stanhopea* orchids produce fewer nuts.

Clearing a wood or filling in a pond could lead to the removal of two guilds of pollinators, which would in turn reduce the diversity of the local flora. A reduction in diversity is always assumed to be 'a bad thing', because sound scientific research has now shown repeatedly that species diversity enhances ecological resilience. In turn, pollinator diversity depends on habitat diversity in order to provide continuity of nectar supply, because most pollinators need to feed for much longer than any single plant species produces nectar. In southern England, for example, woodland plants such as bluebells (*Hyacinthoides non-scripta*) are most important to pollinators from February to April, while from May to July meadow plants take over in importance. In August and September, pond and marginal plants flower more profusely, before the meadows come back into flower in October. Put simply, diversity begets diversity – and the reverse is equally true. If you reduce species and/or habitat diversity, the habitats will become increasingly liable to instability.

SCENT SHOPPING

Stanhopea orchids are pollinated by male bees, with the flowers providing a fragrance that attracts females.

From a research point of view, the role of specific pollinators is difficult to investigate because their distributions and relative proportions vary with geographical locations. Researching biology can feel like trying to hit a moving target, which is why there can be contradictory results in the scientific literature. For example, if you look at the pollination of European plants, 25 per cent of the species are wind pollinated and the remaining 75 per cent are pollinated by invertebrates. In contrast, in the biodiversity hotspot of Western Australia, only 8 per cent of the species are wind pollinated, whereas the same number (75 per cent) are pollinated by invertebrates; the remaining 17 per cent are either pollinated by birds (15 per cent) or mammals (2 per cent). These differences in the relative proportions of the pollinators are of profound ecological importance because the different pollinators require a different landscape, and the landscape is different because of the pollinators.

BEES OF BRAZIL

In order to produce a bumper crop, the flowers of Brazil nut trees (*Bertholletia excelsa*) must be pollinated by female bees.

SEASONAL SUPPLY

Bluebells (*Hyacinthoides non-scripta*) feed pollinators in spring, but without summer flowers, those pollinators would starve.

The benefits of bats

Historians make analysis of the past manageable by looking at a topic through a particular lens. The same trick can be applied to natural history and is used here to consider ecosystem services through the lens of bats and bat pollination.

Apart from the fact that they are nocturnal, bats make good study subjects because they are relatively large, there are relatively few bat species and we have a complete catalogue of them, and bat pollination is one of the more straightforward pollination syndromes (see Chapter 6). The value of bat pollination was comprehensively reviewed in 2011 by a group led by biologist Thomas Kunz. Pollination is just one of the services provided by bats (the others being seed dispersal and crop pest control), and two specific habitats would be very different and depauperate in the absence of this.

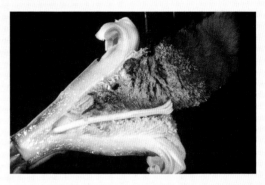

THE CAP FITS
Saguaro flowers 'fit' the heads of bats due to many generations of evolution. If pollinators disappeared and saguaro declined, it would have a catastrophic effect on the numerous creatures that rely on saguaro.

DESERTS DIMINISHED
Saguaro cacti (*Carnegiea gigantea*) rely on bats for pollination; without bats, there would be no saguaro.

The arid habitats in the New World tropics would be unrecognisable without bat pollination – for a start, keystone species like the bat-pollinated agaves (*Agave* spp.) and cacti would be absent. Another group of bat-pollinated plants in such habitats are species in the Bombacoideae subfamily of the mallow family (Malvaceae). These mallows are important forest trees in many tropical parts of the world and can be the dominant species. In light of their influence, pollinators could justifiably be described as ecosystem facilitators, in the same way that ants are often labelled as ecosystem engineers because of the influence they have on the structures of ecosystems.

Bat-pollinated plants do not just have ecological value. The African baobab (*Adansonia digitata*), for example, has huge cultural significance throughout the continent. Economically important plants pollinated by bats include kapok (*Ceiba pentandra*), cacti, durian (*Durio zibethinus*) and balsa (*Ochroma pyramidale*). However, perhaps the most important plants in this group are the agaves, which give us sisal fibres, mescal and tequila.

The economic value of pollinators

Wild and managed pollinators are essential for the maintenance of global food supplies (see box). The production of animal-pollinated crops has increased threefold in the past 50 years, yet during the same period there have been increasing problems with pollinator declines among both managed and wild pollinators. Wind-pollinated crops are not at risk from pollinator fluctuations in the same way, but no one wants to live on a diet limited to cereals.

The value of a single pollinator

With enough data and time, it is possible to calculate the monetary value of pollinators themselves, as opposed to the value of the crops they are involved with. In 1997, ecologist James Cane did just that, in a study on the southeastern blueberry bee (*Habropoda laboriosa*). Taking into account many factors, including the daily work schedules of the bees, their efficiency as pollinators and their lifespan, Cane discovered that in her lifetime a female bee visits up to 50,000 flowers. He also found that 620 flower visits are needed to supply one nest cell with its provender. From these figures, Cane estimated that during this foraging the same female's work would result in 6,000 ripe marketable blueberries, with a market value of US$120. And that is just one individual bee.

A COUPLE OF KEY FACTS

Seventy-five per cent of food crops depend on animal pollinators and these crops provide 35 per cent of the volume of food consumed by humans

In 2015, it was estimated that animal pollination resulted in crops with a global value of up to US$577 million. Animal-pollinated plants provide the seeds, nuts, fruits, oils and vegetables that supply us with most of the vitamins, micronutrients and minerals we need as part of our diet. Included in animal-pollinated crops are coffee, cocoa and cotton, which are particularly important for the economies of developing countries. The production values of crops that depend on animal pollination are, on average, five times more than those that are pollinated by wind.

Each year, 81 million hives of honeybees produce 1.6 billion kg (3.5 billion lb) of honey with an eye-watering global market value of US$26.4 billion

The health benefits of honey are well documented and discussed, but that made from the flowers of the mānuka bush (*Leptospermum scoparium*) is widely regarded as particularly special and sells for more than £700 per kilogram (£1,500 per pound). The presence of the antibiotic methylglyoxal in the honey is thought to be the source of some of its healing properties. These include aiding wound healing, clearing up a sore throat, treating acne, and perhaps reducing the symptoms of gastric ulcers, irritable bowel syndrome and cystic fibrosis.

MĀNUKA = MONEY
Honey produced by bees feeding on the New Zealand native shrub mānuka (*Leptospermum scoparium*) is highly valued.

The size and identity of the workforce

Any investigation into the monetary value of pollinators' activities is not complete without consideration of the production of coffee. Coffee is a major globally traded commodity, but it is often grown in developing countries by small subsistence farmers. The global production of coffee in 2016 was officially recorded as nearly 152 million 60 kg (130 lb) bags. If an average bean weighs 0.1 g (0.004 oz), then there are 600,000 beans in one bag and the annual production of coffee amounts to more than 90 trillion beans! It is thought that about 50 per cent of the flowers of coffee plants (*Coffea* species and hybrids) self-fertilise, and also that one visit from a pollinator is enough to pollinate both ovules in a flower. Despite these provisos, the global coffee crop is the result of 22.5 trillion flower visits. Taking the figure above that a bee makes 50,000 flower visits during its lifetime, the coffee crop therefore 'employs' 450 million bees and other pollinators. It has also been shown that highland coffee crops produce greater yields if a variety of pollinators are doing the work. Yet again, here is evidence that diversity is an essential component of biological systems.

An important point to note here is that crops are often pollinated by a mixture of wild and managed populations of pollinators. The six most valuable crops in the US that are pollinated by at least some native pollinators are (in descending order) soybean, cotton, grapes, apples, oranges and almonds. The pollination of the flowers of soybean plants (*Glycine max*) is carried out by equal numbers of native and domesticated bees, whereas 90 per cent of oranges are produced following pollination by managed bees (which make delicious honey as a result). On the other hand, 90 per cent of the pollination of grape flowers (*Vitis vinifera* cultivars) is by populations of native bees.

The economic cost of pollination

While the economic benefits of pollination are reasonably clear, there are also some associated costs. The flowers of cotton plants (*Gossypium* species and hybrids) in the US are pollinated predominantly by managed populations of bees. While cotton is still a very important crop in economic terms, the history of the industry does not make for easy reading. While the production of cotton did make some people very rich, its cost in terms of human suffering is incalculable, both in relation to the exploitation of slaves and, in turn, its role in setting off the American Civil War. Even today, cotton production remains highly controversial, in terms of its use of water, pesticides and child labour.

There is another indirect but not insignificant cost to humans that is related to pollination: hay fever, or seasonal allergic rhinitis. The economic burden of hay fever comes under several headings. First, there is the cost of medication to reduce the symptoms or prevent a reaction. In countries with a national health service, hay fever patients will be a drain on the resources of that service. A second cost to the economy is the number of adult working days lost each year. In the US and UK, this runs into millions of days each year. Third, there is absenteeism from school and the effect that has on the quality of the education and subsequent value of the students in the workforce later in life. Finally, there is simply the misery caused by hay fever, and the fact that it can make other conditions such as asthma worse and potentially dangerous to health. There have been so many studies into these costs that there are now studies into the ways the costs are calculated.

POLLINATOR PRODUCE
A wide range of pollinators translates into a bumper crop of coffee beans.

TOP CROPS

The six most valuable US crops that rely on wild animal pollinators are soybean (*Glycine max*, top left), cotton (*Gossypium* spp., top right), grapes (*Vitis vinifera*, middle left), apples (*Malus domestica*, middle right), oranges (*Citrus* spp., bottom left) and sweet almonds (*Prunus dulcis*, bottom right).

Putting a price on nature

There is no doubt in the minds of many people – and not just farmers, gardeners and scientists – that pollinators are important for the economic benefits their work brings. As seen above, it is possible to place a figure on the value of pollination to crops such as coffee, and calculating the value of honey is even more straightforward. However, placing a global figure on the value of pollinators to agricultural and horticultural production has proved to be very difficult. This is partly due to the fact that the crops vary in the proportion of their yield that results from wild pollinators. Furthermore, there is great geographical variation in the prevalence of wild pollinators. Despite all these variations, there is a consensus in the scientific literature that wild pollinators are important to agricultural yields.

But can a price be put on the services that pollinators and pollination provide to nature? How important are pollinators to the sustainable functioning of nature in monetary terms? In 1997, a paper published in the journal *Nature* started a movement that has led to a great deal of heated argument – and even the creation of new journals to publish the resulting papers. In it, ecological economist Robert Costanza and colleagues calculated a 'value of the world's ecosystem services and natural capital' to humans, putting this figure at an annual average of US$33 trillion. This was pounced on by many people, especially politicians, and the paper has been cited in more than 4,000 subsequent papers. In 2014, the value of natural capital was recalculated, and it came out at US$125 trillion, but this increase hid a loss of US$20 trillion worth of services. Even people who passionately oppose the principle of natural capital valuations agree that things are getting worse for nature and for pollinators. Pollination was one of the 17 ecosystem services priced by the Costanza team, and its inclusion alone is a very important admission of the need to look after pollinators. The figure they placed on pollination was US$117 billion.

The original paper was greeted with a storm of criticism from economists and biologists alike. A major difficulty with the idea of placing a monetary value on natural resources is that they change with time and context. For example, rarity does not increase monetary value – for example, if an area of desert dominated by bat-pollinated cacti and agaves halves, the value of the bats as pollinators halves. A similar situation would arise if a new pesticide was introduced that controlled pests formerly eaten by bats.

Another simple criticism is that some authors have attempted to value insect pollination services by estimating what it would cost to replace them. The estimates are based on introducing hand pollination or pollen dusting, but these techniques are not applicable to non-commercial habitats. For some ecosystem

REPLACING NATURE

Pollinating commercial vanilla (*Vanilla planifolia*) is done by hand, as its original natural pollinator does not occur in most growing areas. This greatly increases production costs.

functions there are no alternatives to the current service providers – they are simply irreplaceable.

Other criticisms of natural capital include concerns that if the value of nature is included in company balance sheets, then the question of ownership will be raised. This could lead to trading in natural resources that are currently considered to be in common ownership, and the so-called tragedy of the commons would come into play. There is doubt that company shareholders are the best people to be making decisions about the future of the natural resources that humans rely on to keep the Earth sustainably habitable. However, research in 2015 showed that a majority of UK taxpayers would be willing to pay £13.40 per annum for the conservation of pollinators through the taxation system, raising the required £379 million. This is similar to the £350 million that UK gardeners currently spend on bird seed each year out of their taxed income.

Despite these criticisms, natural capital has been accepted as a serious way of examining nature. It appears that politicians do accept a price tag as a reason to conserve biology because natural capital is mentioned in the current UK 25-year plan for the environment. If policy makers are positively engaged in the conservation of our natural resources, then there is a better chance that conservation targets will be met and that thinking around natural capital will have been beneficial despite its flaws, faults and failings. As more and more research, synthesis and reviews are published, it appears that the economists are discovering that Schrödinger was right, and that economic theory is difficult to apply to a system as complicated as nature, with its countless interconnections. We live in interesting times.

Chapter summary

Pollinators are extremely valuable to both natural systems and production systems. Pollinators are a vital and irreplaceable component of nature as we know it today. The conservation, protection and rehabilitation of genetic, species and ecological diversity will succeed only if pollinators are present. This conservation, protection and rehabilitation is the subject of the final chapter.

Chapter 8

Conservation of Pollinators

Pollination is a vital component in the process of life on Earth. Without it, nature will not be able to continue in the forward direction in which it is travelling. There can be no more doubting the need for pollinators if humans are to be fed adequately, and if natural ecosystems and their functions and services are to be maintained sustainably. Given that pollinators are regarded as essential, it may come as a surprise that they still need to be protected – and yet there is clear evidence that they are declining.

Honeybees (*Apis mellifera*) flock to hyacinth (*Hyacinthus* sp.) flowers.

Mismatched needs

The sudden collapse of managed populations of honeybees (*Apis mellifera*) has been very widely reported, as have the various controversial causes behind them. As a result, research has tended to focus more on the conservation of pollinators in human-dominated landscapes than on those in the wild. While this is understandable, it conserves only pollination services (which by definition are for humans) and not pollination functions or pollinator diversity (which are for nature as a whole). In short, research into pollinator declines in human-dominated landscapes focuses on individual species, while that into pollinator declines in the wild takes a whole ecosystem approach.

The demise of the natural populations of some of the 350,000 pollinator species is recognised in the upper echelons of the conservation community and even at government level, but the recognition of a problem is not the same as the solution to the problem. When 15 different international initiatives for large-scale conservation were examined by ecologist Irawan Asaad and co-workers in 2017, not one of them specifically named pollinators.

HELPING HONEYBEES
Although cultivated honeybees are in decline, we should not forget wild pollinator populations.

Are pollinators too easily forgotten?

It can be claimed that pollinators have been ignored in favour of charismatic wet-nosed mega-vertebrates and birds. This might not be entirely fair, because in order to conserve animals such as birds in their habitats, plants are required, and these in turn require pollinators if they are to regenerate, either annually or in the longer term. There is a cost implication too. Research in 2019 showed that animals cost 8–26 times more to conserve than plants, and 13–19 times more than aquatic invertebrates. Bird species can cost 5–30 times more than a plant species to conserve, and 6–14 times more than an aquatic invertebrate species.

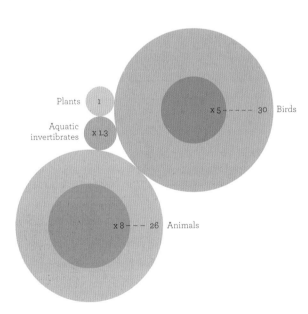

COUNTING THE COST
In general, conserving animals costs much more than conserving plants (drawn using data from Gordon *et al.* 2019).

Conserve and protect a bird, however, and you conserve and protect everything below it in the food chain and the web of life. Likewise, it could be said that if we can show that pollinators are thriving then the ecosystem is functioning well. But are they thriving? The answer to this is 'not everywhere'. On 12 September 2019, a passionate letter was published in *Nature* pleading that the Colombian government should honour its commitment to protect pollinators. One of the problems here is the widespread use of glyphosate to kill cocaine plants, and deforestation to provide land for subsistence farming is not helping the situation either. No one said that these decisions are easy.

Is there really a problem?

In December 2016, the Conference of the Parties to the Convention on Biological Diversity encouraged the 180 governments present to increase their efforts to improve the conservation and management of pollinators, to address the drivers of pollinator declines, and to work towards sustainable food production systems and agriculture. This sentiment was repeated almost two years later at the next meeting. The catalyst for this initiative was the 556-page assessment on the state of pollinators published in 2016 by the Intergovernmental Science-Policy Platform on Biodiversity and Ecosystem Services (IPBES; see box on page 186).

Some evidence

There is no doubt in the minds of the IPBES report authors that wild pollinators have declined in occurrence, diversity and abundance in northwest Europe and North America. However, this is not the same as saying that the rest of the world is doing fine; rather, the problem is that there is a lack of data for the rest of the world. The highest-quality data come from the UK, simply because it has the most comprehensive biological records in the world. And some of these data paint a very disturbing picture. A 2014 paper by biodiversity ecologist Jeff Ollerton and colleagues showed that there were 23 extinctions of bees and flower-visiting wasps between

1853 and 1990 in the UK. The reasons for these extinctions are common to all developed nations with intensive agriculture, so it is fair to assume that the pollinators in all these countries have declined in the same way and to the same extent.

Another study, published in 2013 and based on research carried out in temperate woodland in Illinois, discovered that 50 per cent of the bee species had become locally extinct, or extirpated, in the past 120 years. The authors attributed these dramatic declines to climate changes that had resulted in a phenological mismatch between the flowering times of the plants and the life histories of the bees. Furthermore, they showed that quality and quantity of pollination services had declined, even though the habitat was still present. The declines have not led to collapse of the woodland community. But the authors point out, as do many other researchers, that relying on the level of redundancy is a solution of last resort. Redundancy is the ecological term used to describe a situation where less abundant species are waiting in the wings to take over a role should one of the major players disappear – it is a Get Out of Jail Free card that can be used only once.

In 2015, a global assessment of bird and mammal pollinators was released. Such an assessment is possible because these are large animals for which we have complete species lists and relatively complete conservation assessments. (In comparison, in 2013 only 152 ant, bee and wasp species had an International Union for Conservation of Nature Red List assessment.) This assessment showed that every year one bird pollinator species and two mammal pollinator species become more threatened with extinction, out of a total of 1,432 vertebrate species that pollinate flowers. If this rate is true of all pollinator groups, then more than 750 species of pollinators decline each year. This is worrying, because as Claire Kremen, Kathy Willis and other conservation biologists have shown repeatedly, reductions in species diversity cause reductions in ecological resilience, just as reductions in genetic diversity cause a reduction in evolution and thus resilience to change. Furthermore, not only does diversity support diversity, it actually promotes more diversity. Another consequence of this need to protect all species means that focusing on generalist pollinator species alone is short-sighted.

FUNCTIONAL FOREST
Although wild forests seem intact, pollinator diversity and effectiveness have reduced due to climate change.

HIVE COLLAPSE
Declines in managed bee colonies harm crop production.

Is the problem real or assumed?

In the interests of fairness, it should be pointed out that there is a view that the decline in managed populations of pollinators for food crops has been far greater than the decline in wild pollinators. The reasons for the collapse of colonies of managed bees are partly different from those causing the decline of wild pollinators, and some argue that the two situations are fundamentally different. Plant ecologist Jaboury Ghazoul and others also point out that the data for wild invertebrate pollinators are woefully incomplete and inadequate, and that extrapolations are potentially very inaccurate. However, even those cautioning against overstating the problem accept that there is no room for complacency on the basis that pollination is such a fundamental ecosystem process that it should not be taken for granted.

CHAINS OF EXTINCTION

The decline of pollinators is not a standalone problem, but has a knock-on effect – what anthropologist Jared Diamond has called 'chains of extinction'. As already discussed, pollinators are only one type of flower visitor; non-pollinators are 'just' herbivores, but their needs are as real as those of the pollinators. If the decline in pollinators is accompanied by a decline in nectar-rich flowers, then the flower visitors will also decline – and this will be followed by a cascade of changes. These chains of extinction are just one of the reasons for the loss of biological diversity at all three levels – ecosystem, species and genetic.

The causes of pollinator losses

If we accept that pollinators are essential for the future comfort and prosperity of humans, and that they are most probably declining at a serious rate, then we should be trying to reverse the situation. A first step in this rescue mission is to identify what is causing the problem. Only once this has been achieved can successful recovery programmes be planned and executed effectively.

In 1984, Jared Diamond proposed that the major threats to nature could be grouped under four broad headings, christened the evil quartet. In addition to the chains of extinction mentioned above, this quartet includes overkill, habitat destruction and fragmentation, and biological invasions. In 2005, entomologist Edward Wilson suggested that the threats could be grouped according to the acronym HIPPO, which stands for habitat destruction and fragmentation, invasions, pollution, population growth and overexploitation. Here, invasions can be of plants, animals, fungi and microbes, and pollution embraces climate changes that result from increased carbon dioxide levels, as well as visible pollution such as plastics in the oceans. It is important to remember that these drivers of decline can occur in combination, turning two sub-lethal problems into one lethal one.

The effects of agriculture

Pollinators are at risk from all of the threats defined by Diamond and Wilson, often in combination. Change leading to a reduction in the total area of a habitat as well as the quality of the remaining habitat is commonly cited at the top of the list of threats. The spread of intensive agriculture is frequently blamed for this loss, but the spread of human settlements is another reason (although it may not be a totally negative change).

The loss of 23 species of UK bees and flower-visiting wasps reported in 2014 (see pages 201–2) was shown to be linked very closely in part to an increase in the availability of nitrogenous fertilisers. The application of nitrogen to hay meadows reduces the selective advantage of leguminous plants, whose flowers are pollinated by invertebrates. The first wave of extinctions was preceded by the start of guano imports in the second half of the nineteenth century and the second wave came soon after the Haber process (used in the production of artificial nitrogenous fertilisers) was invented in the early twentieth century.

The third wave of extinctions coincided with a decrease in hay meadows as a result of the demise of mixed farming in the 1970s following changes in government agricultural policies and subsidies. At the same time, there was a wholesale uprooting of hedges, which removed important sources of nectar and pollen, as well as foraging and nesting sites for birds. A fourth wave seems to have been caused by the introduction of new pesticides, such as pyrethroids, neonicotinoids and EBI fungicides (although this link is disputed by the manufacturers of these products). The slowdown in the loss of species in recent years is thought to be due to the fact that all the vulnerable species have already been extirpated.

GROWING PROBLEM
Destruction of habitat for agricultural expansion contributes to pollinator loss.

FOREST FRAGMENTS
Interspersing fields between pockets of forest reduces pollinator diversity.

Habitat fragmentation

Habitat fragmentation is a problem for both the pollinators and the plants they pollinate. In fact, pollination limitation is the actual reason that plant communities dwindle in fragmented habitats. This can be very serious for plants that are self-incompatible and in which the *S*-locus allele system operates (see page 41). As the population size is reduced, the likelihood of all the plants having the same *S*-locus allele increases. This was the problem faced by the Australian button wrinklewort (*Rutidosis leptorrhynchoides*). Following fragmentation of the species' natural grassland and woodland habitat in southeastern Australia, populations were numerous but were all too small to be able to produce enough viable seed because all the plants had the same incompatibility allele. Moving plants can be a temporary solution to this problem.

For pollinators that cannot fly very far, fragmentation is a particularly serious problem. They may suffer from a lack of sufficient food as the plants they rely on become widely dispersed, and they may also face difficulties in finding nesting sites. As a result, their population declines and there is an increase in inbreeding of both the pollinators and the plants they pollinate. Bumblebees are one such group of pollinators, whose social nature and monogamous breeding system make inbreeding a likely result of habitat fragmentation.

MONOGAMOUS RELATIONSHIP
Cowled friar orchid (*Pterogodium catholicum*) relies entirely on one bee species for pollination, so the minimum area of habitat required is greater than that of another plant with multiple pollinators.

DIVIDE AND CONQUER
When the habitat of Australian button wrinklewort (*Rutidosis leptorrhynchoides*) was divided, seed production declined.

A common solution for habitat fragmentation is to create corridors along which the pollinators can travel. The minimum area for a plant and/or its pollinators is species specific – in the case of the South African cowled friar orchid (*Pterygodium catholicum*) and its specialist bee pollinator *Rediviva peringueyi*, for example, the minimum area for a viable population of both species is 385 ha (951 acres), which is equivalent to nearly 200 football pitches.

Biological invasions

Bees and other pollinators have suffered from biological invasions of two types. First, there are invasions of pathogens and parasites. These include the well-documented ectoparasitic *Varroa* mite that affects social bees, and introduced protozoan parasites and tracheal mites found on North American bumblebees. The mites arrived in colonies used to pollinate greenhouse crops, but they have now spilled over into wild populations. Honeybees have also been identified as carriers for small parasitic hive beetles and viruses such as deformed wing virus.

Second, invasive animal and plant species have affected pollinators. Introduced bee species can outcompete native bees, and if the newcomers visit flowers for nectar and pollen but do not effect pollination, their presence is doubly harmful. This situation arose in South America when the native Patagonian giant bumblebee (*Bombus dahlbomii*) was displaced by competition from the introduced European buff-tailed bumblebee (*B. terrestris*) and the large garden bumblebee (*B. ruderatus*). As a result of this, the local raspberry crop suffered because of damage to the flowers' styles through overenthusiastic visitation by the European species.

ALIEN INVADER
European buff-tailed bumblebees (*Bombus terrestris*) are widely used to pollinate greenhouse crops but frequently escape.

The best evidence for the potential of invasions to cause serious problems comes from oceanic islands. Native bees in Hawai'i, for example, were driven to extinction by introduced western yellowjacket wasps (*Vespula pensylvanica*) and Caroline anole lizards (*Anolis carolinensis*), and as a result introduced honeybees are now essential for the pollination of some of the native flora.

GIANT KILLER
A major pollinator of native plants in southern South America, giant bumblebees (*Bombus dahlbomii*) have greatly decreased in number since two non-native bumblebees were introduced into their range. The invaders compete for food and have brought with them new pathogens.

MITE ATTACK

Varroa destructor mites (top left on the back of the bee) are common pests of domestic honeybees. Originating in Asia, they arrived in Europe in the 1960s, South America in the 1970s and North America in the 1980s. Only Australia and Hawaii are said to be varroa-free.

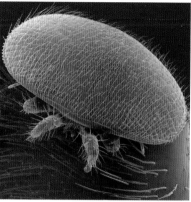

DESTRUCTOR IN DETAIL

Button-shaped *Varroa* mites can breed only in honeybee colonies. They latch onto bees and feed on the 'fat body', an important structure that stores fats and regulates the endocrine system. Mites are also responsible for introducing pathogenic viruses into the colony and can ultimately cause it to perish.

BALSAM INVASION

A well-known example of pollination disruption by an invasive species in the UK involves the Himalayan balsam (*Impatiens balsamifera*), which has widely colonised the banks of streams and rivers, replacing native perennial species in the process.

This is an annual species, and so in winter, when the rivers and streams are flowing at their fastest, there are fewer roots to hold the soil in place, leading to bank erosion and reduced water quality. But this is not the only problem.

The flowers of Himalayan balsam are very rich in nectar and UK pollinators love them. Possible consequences of this are varied. Pollinators may ignore native plants in favour of the balsam. Alternatively, the balsam may be acting as a magnet species and more pollinators may be attracted to the area, resulting in more pollinators for all species. On the other hand, the pollen from the balsam might clog up the stigmatic surfaces of the other species. Then again, native species that do not share pollinators with the balsam might start to increase in number more rapidly than those species that are pollinated by the same animals as the balsam.

Given the complexity of the situation, it is perhaps not surprising that the effects of Himalayan balsam on native pollinators and pollination are unclear. The success of the balsam supports the idea that plants pollinated by generalist pollinators stand a better chance of naturalising in a new region. However, other research shows that self-compatible plants do, in fact, make up a significantly large proportion of exotic species around the world.

It is important to remember that not all non-native species have a negative effect all of the time. In New Zealand, for example, the kiekie (*Freycinetia baueriana*) is pollinated by bats, one of which is missing, presumed extinct, while the other is becoming rarer. However, the plant continues to produce viable seeds in well-developed fruits. It seems that the introduced brushtail possum (*Trichosurus vulpecula*) is not only pollinating the flowers but also dispersing the fruits. That said, the possum is having a devastating effect on many other native plants and bird species.

The threats posed by biological invasions are often identified only when it is too late. By definition, invasions are where an organism is damaging the ecosystem and it cannot be controlled. They have been successfully repelled only on oceanic islands, which ironically and for reasons we do not fully understand are more prone to invasion. Preventing the movement of organisms around the world is the only way to prevent more invasions, but that is not to say that trying to reduce the impact of invasive species is pointless. Clearance of the invasive Port Jackson willow (*Acacia saligna*) in South Africa, for example, has led to the recovery of a damselfly and dragonflies, which have been taken off the possibly extinct list following clearance of the willows and pond restoration. Although these insects are not pollinators, their resurrection shows that all is not lost.

Pollution and climate change

Intensive agriculture involves the use of large quantities of insecticides, which directly affect pollinators, and herbicides, which remove a potential food supply for pollinators. It has been argued that the introduction of cotton varieties that are genetically modified to be resistant to stem-boring caterpillars has reduced the high volume of pesticides poured onto this crop each year. Sadly, however, a report published in 2019 showed that in the 22 years to 2014, the pesticides used in the US became more toxic, not less so. The primary responsibility for this increase was accounted for by the neonicotinoids being sprayed onto corn and soybean crops. In 2018, the European Union banned the use of the three main neonicotinoids on field crops because of their impact on pollinators.

The effects of increased carbon dioxide in the atmosphere, along with other greenhouse gases, are well documented and widely accepted. When the climate changes, there are four possible responses from organisms: they may migrate, they may evolve (adapt), they may tolerate (in other words, exhibit phenotypic plasticity) or they may become extinct (locally or globally). All of these options are available to plants and their pollinators, and while their responses may not be coordinated, they can be. Wytham Woods near Oxford has been monitored for decades, allowing scientists to research the effects of early, average and late springs. The relationship between the great tits (*Parus major*) here, the invertebrates they feed to their chicks and the plants on which the invertebrates feed remains intact because all three elements react in synchrony.

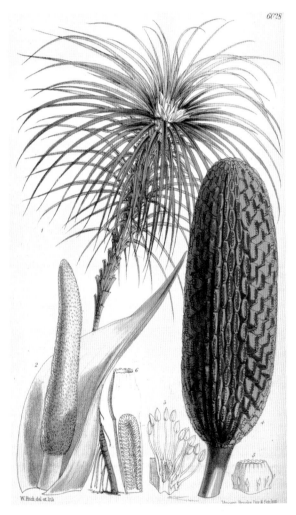

KIEKIE
Freycinetia baueriana.

FLY SPRAY
Widespread use of pesticides in commercial farming kills both pests and potential pollinators.

Rising temperatures can affect pollination by changing the phenology of the organisms involved, but it can also alter the geographical ranges of the species. If the ranges of the pollinator and plant species no longer overlap, then the plant has to hope that another species will step up and pollinate its flowers. Meanwhile, the pollinator has to look for a new species to feed on, and hopefully in the process it also pollinates the flowers of that species. Clearly, this suggests that plants with generalist flowers pollinated by a wide range of visitors, and generalist pollinators capable of feeding from many flowers, will be at an advantage over plants and animals in one-to-one specialist relationships.

More than one review has suggested that climate change will lead to a depauperate world of generalists. This is important, because work by agroecologist Heather Grab and colleagues published in 2019 shows that in order to maintain optimum fruit harvests in apple orchards (in terms of fruit weight and quality, and seeds within the fruit), it is not enough to have an abundance of pollinators. Those pollinators also need to be from many unrelated species. And what applies to apples may also be true for wild plants in their natural habitats.

Pollution does not have to be chemical. For example, light pollution in urban areas can be disruptive to the behaviour of moths, whether they are diurnal, nocturnal or crepuscular. When artificial light reaches certain levels, it leads to reduced activity in nocturnal moths.

And while the diurnal and crepuscular moths may remain active for longer as a result of the light, they do not provide the same services as the nocturnal species. These moths stick to the flowers they pollinate in natural conditions and do not also visit the flowers pollinated by nocturnal moths.

LIGHT AT NIGHT
Although lights tend to attract night-flying insects, light pollution seems to reduce activity in moths, which are important pollinators for many plants.

THE CAUSES OF POLLINATOR LOSSES

Solutions to pollinator loss and decline

Conservation activity tends to focus on one of three levels: genetic, species and ecosystem. However, while a project may focus on just one level, the other two will both be important components. For example, the species recovery programme for the western underground orchid (*Rhizanthella gardneri*) in Western Australia states that fungus gnats pollinate the flowers, but as there are plenty of these insects they are not seen as a factor likely to limit the success of the project. Instead, lack of seed dispersal, weed invasion and inappropriate fire management are all recognised as more important.

Similarly, conservation activity tends to focus on either animals or plants, and less commonly on both in equal measure. The Global Strategy for Plant Conservation, adopted by the Conference of the Parties to the Convention on Biological Diversity in 2002, was the first global programme for any major group of organisms. Strangely, however, it fails to include either explicit mention of pollinators or a target for their conservation. In 2015, entomologist Carl Wardhaugh made a plea for 'greater emphasis on the entomological, rather than the botanical, side of the flower–invertebrate interaction', but this will happen only when botanists and entomologists start to interact as closely as pollinators and their flowers, or flowers and their pollinators.

Pollinators are not forgotten

Despite Wardhaugh's criticism, pollinators are considered in conservation programmes. For example, one of the performance indicators of the restoration of traditional wildflower (MG5) meadows at Harcourt Arboretum in Oxfordshire was the establishment of lesser knapweed (*Centaurea nigra*), because it provides valuable nectar to pollinators late in the summer and early autumn. Furthermore, the project involved the creation of new ponds to ensure continuity of nectar from early spring right through to autumn. In addition, it included the planting of kilometres of hedges containing 12 different native species, autumn grazing by cattle and sheep, and hay removal to maintain the sward diversity and promote legumes. All these actions have been shown to promote bumblebees in particular. Another improvement was the provision of tussock-forming grasses around the margins of the meadow and in fenced-off areas. These sites are the chosen habitat of small mammals, whose abandoned holes are used by bees for nests.

POLLINATOR CONTINUITY
An abundance of pollinators such as the common brimstone (*Gonepteryx rhamni*) suggests a healthy diversity of flowers.

In a much larger habitat restoration project in southwest Western Australia, pollinators are seen as pivotal to its success. The region is a global biodiversity hotspot, and in such areas there is always a high level of specialised pollination syndromes. Botanist Kingsley Dixon, who has been an inspirational leader in plant conservation for more than three decades, has been particularly important in furthering our understanding of the role pollinators play in habitat rehabilitation. Dixon works mostly in Western Australia and his mantra is that 'animal-based pollination services sustain the reproductive potential and genetic resilience of ecosystems'. While specialist pollinators are the most likely pollinators to become extinct, the loss of generalists will have a more pervasive impact.

It has been shown that pollinators (both specialists and generalists) with a low dispersal potential are particularly at risk from current drivers of decline. In order to support them, larger areas in the landscape have to be protected and then joined by corridors. Successful corridors must contain framework species that provide a major amount of nectar and pollen, bridging species that provide resources in otherwise

MOTORWAY NETWORK
By providing habitat corridors, pollinators can move between protected areas.

quiet times, and magnet species that have attractive flowers and bring in the pollinators for the drab flowers that would otherwise be missed. In summary, the success of landscape-scale restoration depends on understanding the dispersibility and migration of pollinators, as well as the plant species that facilitate and assist pollinator movement.

The provision of a network of protected areas joined by corridors is not impossible even in an intensive agricultural system such as that in central England. Coppices, ponds, hedgerows and headlands can all be created and will start to have a positive impact within a few years. Agri-environmental schemes have been funded in Europe for many years and have had a positive impact in places, particularly when more than 7.5 per cent of the farmland is left uncropped. Only then can the positive effects of pesticide reduction and organic farming start to have an effect.

Plants are easier to conserve

One reason for the perception that plants get more attention than pollinators is that they are much easier to work with than animals. Take, for example, the case of the cycad *Encephalartos ferox*. This threatened species is pollinated by a beetle that is probably extinct. Like all cycads it is dioecious, and like many cycads it is relatively easy to cultivate and is commonly seen in botanical glasshouses in temperate collections. When the female plant at Oxford Botanic Garden produced a cone and pollen for it was donated from the Royal Botanic Gardens, Kew, that pollen had to be forced into the cone without the services of the beetle. Stewart Henchie from Kew devised a workaround. He put the pollen in water, sucked the solution up into his bicycle pump, attached the flexible valve hose and then squirted the pollen deep into the cone. This worked at Kew, but by the time the Oxford plant needed to be pollinated Henchie had improved the system, replacing the bicycle pump with a turkey baster because this makes a better surrogate for the beetle.

When species recovery programmes are compared, pollination requirements can be very different. *Euphorbia stygiana* is a shrubby perennial that is endemic to the Azores. Pollination is not a problem because spurges rarely need a specific pollinator – they are successfully pollinated by a wide range of vectors and have high levels of self-compatibility. In contrast, the dune thistle (*Cirsium pitcheri*), a species native to sand dunes bordering the North American Great Lakes, is monocarpic and dies after three to five years. Fortunately, this is a generalist plant species and is self-compatible, so pollination is not high on the list of problems it faces. The same cannot be said of Mead's milkweed (*Asclepias meadii*), a native of the Midwest prairies of the US. This is a long-lived perennial, but it is self-incompatible and outcrossing is essential – different genotypes are required if the plants are to reproduce. Finally, the prairie white fringed orchid (*Platanthera leucophaea*) requires the services of a hawkmoth pollinator, in the absence of which hand pollination is the only alternative.

So plants can be conserved much more easily than animals, large or small, but can they be put back in the wild and form self-propagating populations? A review in 2019 by ecologist Thomas Abeli and co-workers showed that of the first 13 attempts to restore an extinct plant species from cultivated material, five completely failed. While a 60 per cent success rate is better than nothing, it does show that there is no substitute for the preservation, restoration or rehabilitation of habitats. Nor is there room for complacency.

POLLINATION NO BAR
The rare dune thistle (*Cirsium pitcheri*) is pollinated by several different insects.

A HELPING HAND
When the right moth is unavailable, prairie white fringed orchid (*Platanthera leucophaea*) must be hand pollinated.

An inspirational success story

Proof that locally extinct insect species can be restored is found in the work and achievements of the inspirational ecologist Jeremy Thomas and his team. Working in the south of England, this group restored the locally extinct large blue butterfly (*Maculinea arion*; now *Phengaris arion*).

This insect feeds on the flowers of thyme species (*Thymus* spp.) in grassland. It lays its eggs on the plants, and when the larvae are partially developed they drop to the ground. They then excrete a sweet fluid that attracts *Myrmica sabuleti* ants, which mistake them for their own grubs and take them back to their nest. Once there, the butterfly larvae feed on the ant grubs, before pupating and emerging from the nest the following spring.

The key factor in this extraordinary relationship is the temperature required by the ants in their nest. If the grass is allowed to grow taller than 14 mm (0.5 in), then the shade cast by the leaves keeps the nest too cool and other ant species become dominant in the area. The original demise of the butterfly was precipitated by fencing that was intended to keep out humans but inadvertently kept out herbivores. As a result, the grass grew too tall, the ants disappeared and, in turn, the large blue butterfly population died out.

This project is now a paradigm for butterfly conservation. It combined meticulous, broad data collection with multidimensional mathematical modelling to suggest a strategy that has worked successfully at more than 78 sites.

ANT ANTICS
Breckland thyme (*Thymus serpyllum*) feeds adult large blues (*Phengaris arion*), but their larvae need help from the right type of ant.

Gardening the world

In 2019, plant ecologist Molly McCormick and colleagues reported on a successful pollinator restoration project in Arizona. They found that the best strategy is to create species-rich patches, or gardens as they call them, that flower abundantly. The idea that gardening is very close to habitat rehabilitation is finally starting to gain traction, partly because gardeners often have an interest in conservation and want to be involved, and partly because they also have the skills necessary to raise and cultivate plants. Most gardeners live in urban environments, and evidence is now revealing that urban gardens will become very important for the conservation of pollinators, just as they are important for songbird conservation.

In 2013, the UK Insect Pollinators Initiative was launched. This multidisciplinary project has many streams of funding, indicating the complexity of the subject, and one specific area that has been investigated is the role of urban pollinators. The interest in cities as refuges for pollinators coincides with a new style of gardening coming out of Sheffield University and led by horticultural ecologists James Hitchmough and Nigel Dunnett. Their strategy is to create ecologically diverse urban vegetation, in which the plantings are more random and often use varieties that are closer to wild plants than modern garden cultivars, some of which no longer produce nectar rewards.

Nor is this idea limited to the UK. In the West Bengal city of Kolkata, the seventh-largest city in India, corridors are being created to join up green areas into a butterfly plant network. An added dimension to this work is that it brings many people into contact with nature, and there is no doubt that citizen engagement with conservation is of critical importance to the success of such projects. The involvement of citizens is now seen as going beyond back gardens and into data collection and other aspects of research.

The future of pollination is in our hands

For communities of pollinators and plants to survive sustainably into the future there must be closer associations between botanists and zoologists, and between conservation workers and the public. On the plus side, as we enter the third decade of the twenty-first century there is a new momentum to conserve pollinators, based on the realisation that in some cases we will face a severe financial penalty if ecosystem functions are not protected. However, there is also a realisation that conservation itself comes at a cost, and evidence that the current failure to hit conservation targets is not due to incorrect paradigms but instead a result of insufficient allocation of financial resources.

There is also a growing call for a review of the relationship between the natural sciences and the humanities. Research workers in both of these fields are calling for paradigm shifts in the culture of conservation, so that it creates landscapes in which both humans and species-rich biological diversity can thrive. It is finally being acknowledged that conservation is successful only when the presence and needs of the billions of human beings are taken into account. The discipline of environmental humanities is at the centre of this new partnership.

GREEN FINGERS
Flower-filled domestic gardens provide an important habitat for wild pollinators.

The climate of the Earth is changing and action is required now. We already have enough information to make conservation effective, but we also need to engage with nature and to develop a sense of belonging and responsibility for the past, the present and the future. There must be trust and collaboration between scientists and citizens, and there must be incentives to change behaviours alongside legislation. In the challenging words of Robert May, Baron May of Oxford, future generations will find it 'blankly incomprehensible' if we do not act now, when we have the knowledge and resources to make a difference.

NEIGHBOURHOOD WATCH
Without the commitment, involvement and understanding of local communities, conservation projects rarely succeed, and the public have never before been more involved in science.

SUNSET STORY
The intricate and beautiful relationships between plants and their pollinators have never been more at threat, so we must work to protect them, for our own benefit as much as theirs.

Bibliography

More information on aspects of this subject can be found in these publications:

Allan, M. 1977. *Darwin and His Flowers: The key to natural selection*. Faber & Faber.

Bronstein, J. L. (*ed*.). 2015 *Mutualism*. Oxford University Press.

Chittka, L. and Thomson, J. D. 2001. *Cognitive Ecology of Pollination: Animal behaviour and floral evolution*. Cambridge University Press.

Christenhusz, M. J. M., Fay, M. F. and Chase, M. W. 2017. *Plants of the World*. Royal Botanic Gardens, Kew, and University of Chicago Press.

De Jong, T. and Klinkhamer, P. 2005. *Evolutionary Ecology of Plant Reproductive Strategies*. Cambridge University Press.

Faegri, K. and van der Pijl, L. 1971. *The Principles of Pollination Ecology*. 2nd edn. Pergamon Press.

Futuyma, D. J. and Kirkpatrick, M. 2017. *Evolution*. 4th edn. Sinauer.

Glover, B. 2014. *Understanding Flowers and Flowering*. 2nd edn. Oxford University Press.

Gould, S. J. 1989. *Wonderful Life: The Burgess Shale and the nature of history*. Vintage.

Harder, L. D. and Barrett, S. C. H. 2006. *Ecology and Evolution of Flowers*. Oxford University Press.

Herrera, C. M. and Pellmyr, O. 2002. *Plant-Animal Interactions: An evolutionary approach*. Blackwell Publishing.

Intergovernmental Science-Policy Platform on Biodiversity and Ecosystem Services (IPBES). 2016. *The Assessment Report on Biodiversity and Ecosystem Services on pollinators, pollination and food production*. Potts, S. G., Imperatriz-Fonseca, V. L. and Ngo, H. T. (*eds*). Secretariat of the Intergovernmental Science-Policy Platform on Biodiversity and Ecosystem Services.

Johnson, S. D. and Schiestl, F. P. 2016. *Floral Mimicry*. Oxford University Press.

Losos, J. 2017. *Improbable Destinies: How predictable is evolution?* Penguin Books.

Mabberley, D. J. 2017. *Mabberley's Plant-Book*. 4th edn. Cambridge University Press.

May, R. 1988. How many species are there on Earth? *Science*, 241(4872): 1441-1449.

Ollerton, J. 2017. Pollinator diversity: distribution, ecological function, and conservation. *Annual Review of Ecology, Evolution, and Systematics*, 48: 353-376.

Patiny, S. (*ed*.) 2012. *Evolution of Plant-Pollinator Relationships*. Cambridge University Press.

Procter, M., Yeo, P. and Lack, A. 1996. *The Natural History of Pollination*. Timber Press.

Schaefer, H. M. and Ruxton, G. D. 2011. *Plant-Animal Communication*. Oxford University Press.

Schoonhoven, L. M., van Loon, J. J. A. and Dicke, M. 2005. *Insect-Plant Biology*. 2nd edn. Oxford University Press.

Subsidiary Body on Scientific, Technical and Technological Advice (SBSTTA). 2018. *The International Pollinator Initiative Plan of Action 2018-2030*. Subsidiary Body on Scientific, Technical and Technological Advice.

Trewavas, A. 2014. *Plant Behaviour and Intelligence*. Oxford University Press.

Waser, N. M. and Ollerton, J. (*eds*). 2006. *Plant-Pollinator Interactions: From specialization to generalization*. University of Chicago Press.

Willis, K. J. and McElwain, J. C. 2013. *The Evolution of Plants*. 2nd edn. Oxford University Press.

Willmer, P. 2011. *Pollination and Floral Ecology*. Princeton University Press.

References

Chapter 2: Wind and Water

Culley, T. M., Weller, S. G. and Sakai, A. K. 2002. The evolution of wind pollination in angiosperms. *Trends in Ecology & Evolution*, 17(8): 361-369.

Friedman, J. and Barrett, S. C. H. 2008. A phylogenetic analysis of the evolution of wind pollination in the angiosperms. *International Journal of Plant Science*, 169(1): 49-58.

Chapter 3: Animals

Waser, N. M. 2007 Pollination by animals. *Encyclopedia of Life Sciences*. doi: 10.1002/9780470015902. a0003163.pub2.

Chapter 4: Attraction

Chittka, L. and Raine, N. 2008. Recognition of flowers by pollinators. *Current Opinion in Plant Biology*, 9(4): 428-435.

Moyroud, E., Wenzel, T., Middleton, T., Rudall, P. J., Banks, H., Reed, A., Mellers, G. et al. 2017. Disorder in convergent floral nanostructures enhances signalling to bees. *Nature*, 550: 469-474.

Raguso, R. A. 2004. Flowers as sensory billboards: progress towards an integrated understanding of floral advertisement. *Current Opinion in Plant Biology*, 7: 434-440.

Schiestl, F. P. and Johnson, S. D. 2013. Pollinator-mediated evolution of floral signals. *Trends in Ecology & Evolution*, 28(5): 307-315.

Chapter 5: Rewards

Parachnowitsch, A. L., Manson, J. S. and Sletvold, N. 2019. Evolutionary ecology of nectar. *Annals of Botany*, 123(2): 247-261.

Chapter 6: The Biological Significance of Pollination

Armbruster, W. S. 2017. The specialization continuum in pollination systems: diversity of concepts and implications for ecology, evolution and conservation. *Functional Ecology*, 31(1): 88-100.

Ollerton, J. 2017. Pollinator diversity: distribution, ecological function, and conservation. *Annual Review of Ecology, Evolution, and Systematics*. 48: 353-376.

Ollerton, J., Liede-Schumann, S., Endress, M. E., Meve, U., Rech, A. R., Shuttleworth, A., Keller, H. A. *et al*. 2019 The diversity and evolution of pollination systems in large plant clades: Apocynaceae as a case study. *Annals of Botany*, 123(2): 311-325.

Rosas-Guerrero, V., Aguilar, R., Martén-Rodríguez, S., Ashworth, L., Lopezaraiza-Mikel, M., Bastida, J. M. and Quesdada, M. 2014. A quantitative review of pollination syndromes: do floral traits predict effective pollinators? *Ecology Letters*, 17(3): 388-400.

Van der Niet, T., Peakall, R. and Johnson, S. D. 2014. Pollinator-driven ecological speciation in plants: new evidence and future perspectives. *Annals of Botany*, 113(2): 199-211.

Chapter 7: Pollination, Pollinators and People

Intergovernmental Science-Policy Platform on Biodiversity and Ecosystem Services (IPBES). 2016. *The Assessment Report on Biodiversity and Ecosystem Services on pollinators, pollination and food production*. Potts, S. G., Imperatriz-Fonseca, V. L. and Ngo, H. T. (*eds*) Secretariat of the Intergovernmental Science-Policy Platform on Biodiversity and Ecosystem Services.

Potts, S. G., Imperatriz-Fonseca, V., Ngo, H. T., Aizen, M. A., Biesmeijer, J. C., Breeze, T. D., Dicke, L. V. *et al*. 2016. Safeguarding pollinators and their values to human well-being. *Nature*, 540: 220-229.

Chapter 8: Conservation of Pollinators

Goulson, D., Nicholls, E., Botías, C. and Rotheray, E. L. (2015). Bee declines driven by combined stress from parasites, pesticides, and lack of flowers. *Science*, 347(6229): 1435.

Menz, M. H., Phillips, R. D., Winfrel, R., Kremen, C., Aizen, M. A., Johnson, S. D. and Dixon, K. W. 2011. Reconnecting plants and pollinators: challenges in the restoration of pollination mutualisms. *Trends in Plant Science*, 16(1): 4-12.

Majewska, A. A. and Altizer, S. 2019. Planting gardens to support insect pollinators. *Conservation Biology*, 34(1): 15-25.

Index

Page numbers in **bold** refer to information contained in captions.

Abeli, Thomas 212
abiotic pollination 46–7
  see also water pollination; wind pollination
Acacia (acacia) 28, 117
  A. saligna 208
acoustic signals 99, 103, **103**
Acrocephalus scirpaceus 130
actinomorphic blooms 105, **105**
Adansonia digitata 192
African baobab *see Adansonia digitata*
Agave (agave) 138, 171, 192, 196
  A. americana 104, **104**
Aglaiocercus kingii **78**
agri-environmental schemes 211
agriculture
  intensive 202, 208, **209**
  negative impact 202, 204, **204**, 208
  organic 211
  subsistence farming 201
Agrius cingulata **121**
Aleppo spurge *see Euphorbia aleppica*
algae 22, 53
Algarve 93, 118
aliphatic compounds 115
alleles 41, 205
  see also S-locus allele system
allelopathy 19
almond, sweet (*Prunus dulcis*) **182**, 194, **195**
Aloe (aloe) 77, 178, **178**
  A. vryheidensis 178, **178**
Alysinae wasp 105
Amazon water lily *see Victoria amazonica*
amber, insect preservation in 85, **85**, 94, **94**, **184**
Amorphophallus titanum 107
Anagyris foetida 77
Ananas comosus 160
Anemone 81
Angelica sylvestris (angelica) 111
angiosperms 9, 17
  and inbreeding 36, 41
  origin 133
  pollen grains of 18, 34–5, **34**
  and speciation 160–5, 181
  and wind pollination 46, 60–2, 64
Angophora 142
Angraecum sesquipedale 172
Anigozanthos flavidus 109, **109**
animal life history **17**
animal pollination 43, 46, 65, 67–95, 183–97, **184**
  bats 170–1
  birds 170, **170**, 171, **171**, 172, 190
  and conservation efforts **188**, 189, 197, 199–215
  economic value of 193–4, 196–7
  features of in self-pollinating plants 38
  and gymnosperms 55–7, 59, 84–5, 89
  and inbreeding 205

invertebrates 190
  mammals 190
  non-flying mammals 72–3, 170
  and pollination syndromes 169–77
  and pollinator function 68–9
  and pollinator shortages 64, **64**
  and speciation 160–4
  and species diversity 161, 162
  vertebrate pollinators 70–80, 169–72
  and wind pollination 28, 46, 47, 54–5, 169, **169**, 176–7
  see also attracting animal pollinators; rewards
annuals 160
Anolis carolinensis 206
antheridia 14, 17, 18
anthers 18, 148
  and animal pollination 79, 91, **109**, 111, 169
  and inbreeding prevention strategies 40, **40**, 41
  and lizard pollination 169
  opening 26–7
  and pollen-only flowers 139
  pores in 27, **27**, 91
  and presentation of pollen 29, **29**
  and scent 121
  and self-pollination 42
  and water pollination 49, 50, 51, 53
  and wind pollination 61
  yuccas 151
anthesis 120
anthocyanins 108
Anthophora 123
Anthornis melanura **171**
antibiotics 93, 193
antioxidants 87
Antirrhinum 175
ants 69, 90, 92–3, **93**, 117, 133, 176, 202, 213
Aphelandra acanthus 101, **101**
Apiaceae (carrot family) 93, 111, 134, 174
Apis 119, 139, 175
  A. mellifera 30, **68**, 83, **90**, 101, 117, **117**, 199
  colony collapse 200, **200**, 203, **203**
  and invasions 206, **207**
Apocrypta bakeri 152
Apocynaceae (dogbane) 162, **162**
apple 194, **195**, 209
aquatic animals, pollination by 49, 51, **51**
aquatic plants 47, 48–51, 54
Aquilegia 105
  A. formosa **132**
Arabidopsis thaliana 33, 37, 42
Araceae (arum family) 39, **39**, 103, 107, 119, 174
Araucaria 33
archegonia 14, 17
Arias, Jorge 148
arid habitats 133, 192
Aristolochia 139, 147
  A. ringens **131**
Armbruster, Scott 162, 176, 177
Arthrochilus 122
arthropods 131
Artocarpus altilis 147
arum family *see* Araceae

Arum maculatum 39, **39**
Asaad, Irawan 200
Asclepiadaceae (milkweeds) 94
Asclepias
  A. incarnata **162**
  A. meadii 212
Asian cave nectar bat *see Eonycteris spelaea*
Asian clematis *see Clematis stans*
Asian skunk cabbage *see Lysichiton camtschatcensis*
Asphodelus albus **92**
Aster (aster) **99**
Asteraceae (daisy family) 29, **29**, 39, 105, 106–7
  see also Aster; *Symphyotrichum*
attracting animal pollinators 95, 97–125
  and communication 97, 98, 99, 122
  and deception strategies 99, 101, **119**, 122–5, **124**, 130, **131**, 139, 142–4
  and signals 99–125, **102**
  the whole package 122–4
Australia 80, 116, 170, 180, 205
  Western 190, 211
Australian banded honeyeater *see Cissomela pectoralis*
Australian button wrinklewort *see Rutidosis leptorrhynchoides*
Australian trigger plants *see Stylidium*
Azores 212

bacteria 20
bacterial infections 142
Bactris 141
balsa *see Ochroma pyramidale*
balsam fig *see Clusia rosea*
banana *see Musa*
Banksia (banksias) 73, 170
  B. coccinea **170**, **180**
bats 70, 73–7, **74–5**, 81, 101, **101**, 124, 165, 208
  and attraction strategies 103, **103**, 106, **106**, 108, 113, 115
  and ecosystem services 192, **192**
  and pollination syndromes 170–1, **171**, 192
  and rewards 136, 138, **138**, 139
  value of 196
bean family *see* Fabaceae
bee orchid *see Ophrys apifera*
bee-pollinated flowers 27, 28, 55
bees 71, 80, 85, 90–2, **91**, 98, 151, 156, 165, 171, **171**, 190, 194
  attracting 100–1, 105–9, 111, 115–16, 175, 178, **178**
  brains of 107
  and the buzz collection method 91, 139
  connection with humanity 188
  and conservation issues 204
  and deception strategies 130, 142–4, **144**, **145**
  euglossine 142, 175, 190, **190**
  extinction 201–2
  and habitat fragmentation 205, **205**
  and invasions 206
  leg 'baskets' of 30, **30**
  long-tongued 100

nests 142, 210
  and pollination syndromes 175, 178
  rewards and 133, 136–7, **136**, 139, **140**, 141–2, 146, 153, 175
  solitary 141
  and species diversity 175, **175**
  vision 107–9, **107–8**, 121
  see also specific types of bee
beetles 80, **83**, 85, 89, **137**, **163**
  attracting 119, 120
  carrion 6
  and conservation efforts 212, **212**
  and pollination syndromes 174, **174**
  and rewards 137, 139, 146
  see also Coleoptera (beetles); *specific beetles*
behavioural ecology 102
bellbird *see Anthornis melanura*
Bencomia caudata 61
Bennettitales 164
Bentham's cornel *see Cornus capitata*
benzaldehyde 120
benzenoids 115
benzyl acetone 102
Bertholletia excelsa 190, **190**
betalains 108
betanin **108**
Betula 61, 62
Bignoniaceae (bignonia family) 135
biodiversity 186
  corridors 205
  declines in 201–2
  hotspots 165, 190, 211
  see also species diversity
biological fitness, reduced 12
biological species concept 157, **157**
biology 155, 156, 162, 166, 185
biotic pollination 46–7
  see also animal pollination
birch *see Betula*
bird of paradise flowers 79
birds 69–70, 73, 76–81, **101**, 171, **180**, 190
  attracting 113
  and conservation efforts 201, 202
  and pollination syndromes 165, 170, **170–1**, 172, 180
  and rewards 136
  see also specific birds
bisexuality 12, 40, 62, 139, 167
biting midges 69, 94
blackcap (Eurasian) *see Sylvia atricapilla*
black hellebore *see Helleborus niger*
black-and-white ruffed lemur *see Varecia variegata*
blanket weed 53
blowflies 94, **124**, 130
blue-faced honeyeater *see Entomyzon cyanotis*
bluebells 190, **190**
body temperature, butterfly/moth 86
Bombacoideae 192
Bombus 139, 205, 206, 210
  B. dahlbomii 206, **206**
  B. ruderatus 206
  B. terrestris **71**, 146, 206, **206**
Boraginaceae (borage) 19
bracken (*Pteridium*) **16**, 133

INDEX 217

bracts 111, 122–3, **123**, 134, **134**, 139
Brassicogethes aeneus **69**
Brazil nut tree see *Bertholletia excelsa*
breadfruit see *Artocarpus altilis*
Breckland thyme see *Thymus serpyllum*
brimstone see *Gonepteryx rhamni*
bristle worms 49, 51
broad-lip bird orchid see *Chiloglottis trapeziformis*
brood chamber provision 147–51
broombush see *Melaleuca uncinata*
Brugmansia sanguinea **136**
brushtail possum see *Trichosurus vulpecula*
bryophytes **12–13**
bucket orchids see *Coryanthes macrantha*; *Coryanthes panamensis*
buff-tailed bumblebee see *Bombus terrestris*
bulbuls see Pycnontidae
bumblebees see *Bombus*
bunchberry see *Cornus canadensis*
burnet moths see Zygaenidae
buttercup see *Ranunculus*
butterflies 17, 86–8, **86**, **88**, 136, 173, **173**, 179, 188, 214
see also specific butterflies
buzz collection method 91, 139

C4 160, **160**
cacao tree see *Theobroma cacao*
Cactaceae (cactus family) 147, 150, **166**, 192, **192**
caffeine 117
Caltha palustris **107**
Calycanthus 139
CAM see crassulacean acid metabolism
Canadian pondweed see *Elodea canadensis*
Cane, James 193
Cape St Vincent, Portugal 118
carbon dioxide emissions 208
Carex 167, **167**
carnation family see Caryophyllaceae
Carnegiea gigantea 138, **138**, 192
Caroline anole lizard see *Anolis carolinensis*
carotenoids 108, **108**
carpel
 and fruit formation 160
 and protandry 39
 and wind pollination 46, 54
carrion beetles 6
carrion flies 174
carrot family see Apiaceae
Caryophyllaceae (carnation family) 93
castor oil plants see *Ricinus communis*
Catasetum 144
 C. osculatum **145**
caterpillars 17
catkins 61, **61**, **129**, 167, **167**
cats 76
Cecidomyiidae fungal gnats 105
Cedar of Lebanon see *Cedrus libani*
Cedrus (cedar) 26, 55
 C. libani **55**

Ceiba pentandra 192
cell membranes 23, 108
Centaurea nigra 210
century plant see *Agave americana*
Cephalotaxus 59, **59**
Ceratophyllaceaea 52
Ceratophyllum 52–3
 C. demersum **52**, **53**
Ceratosolen solmsi 94, 152, **152**
Cercartetus concinnus **170**
Cerceris arenaria **135**
cereals 193
Cereus repandus 166
Cetonia aurata 89, **89**
Chalcidoidea (fig wasps) 94, 147
Chamaedorea 46–7
 C. elegans 46, **46**
Chamaerops humilis 118, **118**
Chara 53
chats see Saxicolinae
chiffchaffs see *Phylloscopus collybita*
Chiloglottis 122
 C. trapeziformis 180
Chinese fighazel see *Sycopsis sinensis*
Chinese orchid see *Dendrobium sinense*
Chinese witch-hazel see *Hamamelis mollis*
Chittka, Lars 107, 162
chlorins **108**
chlorophyll 108, **108**
Christmas rose see *Helleborus niger*
Christ's tears tree see *Erythrina crista-galli*
cinnabar moth see *Tyria jacobaeae*
cinquefoils see *Potentilla*
Cirsium
 C. pitcheri 212, **212**
 C. rivulare 146
Cissomela pectoralis **67**
Cistaceae 32
Cistus 139
 C. monspeliensis **174**
Citrus **195**
Classopollis **10**
Clematis stans 100, **100**
Clianthus puniceus 42, **42**, 171, **171**
climate change **202**, 204, 208–9, 215
Clusia rosea 142
cluster-head sugarbush see *Protea welwitschii*
co-adaptation 138
co-diversification 164
co-evolution 159, 160
co-radiation 180
Cobaea scandens 103
cocaine 201
Coffea (coffee) 193, 194, **194**, 196
Coleoptera (beetles) **83**, 89
 see also beetles
collagen 138
colour
 green 109, **109**
 leaves 111
 nectar 113, 178
 pollen grains 100, **100**, 101, 111
 ultraviolet 92, **92**, **107**, 109, 111, 121
 see also flower colour
colour vision 87, 92
columbine see *Aquilegia*
common eelgrass see *Zostera marina*

common hazel see *Corylus avellana*
common spotted orchid see *Dactylorhiza fuchsii*
communication 97, 98, 99
competitive exclusion 176
conifers 56
Conopidae 184
conservation **188**, 189, 197, 199–215
 and causes of decline 6, 204–9
 costs of 197, 201, **201**, 214
 and mismatched needs 200
 solutions 210
Convention on Biological Diversity 186, 201, 210
convergent evolution 50, 133, 166, 179–81
corn 208
Cornus
 C. canadensis 28, **28**, 111
 C. capitata 123, **123**
Corollina **10**
corona 121
coronene **115**
corpse lily see *Rafflesia arnoldii*
corridors 205, 211, 214
Coryanthes
 C. macrantha 144, **145**
 C. panamensis **145**
Corylus avellana 61, **61**, 62, **129**, 167
Corymbia 142
Costanza, Robert 196
cotton (*Gossypium*) 193, 194, **195**, 208
cowled friar orchid see *Pterogodium catholicum*
Cox, Paul 49
crab spiders see Thomisidae
crassulacean acid metabolism (CAM) 160, **160**
creeping dogwood see *Cornus canadensis*
crepuscular species 209
Cretaceous period 64
Crete 169
crops 203, 206, 208
 animal-pollinated 193, 194, 196
 genetically modified 208
 wind-pollinated 193
crustaceans 49, 51
cuckoo (European) see *Cuculus canorus*
cuckoo wasps 94, **94**
Cuculus canorus 130
Cullenia exarillata 73
cup and saucer vine see *Cobaea scandens*
cuticles 12, **12**
Cyanomitra olivacea 79, **79**
Cycadothrips 116
cycads 55, 57, **84**, 85, **85**, 116, 133, 147, 164, 167
 see also *Encephalartos ferox*
Cycas revoluta **84**
Cyclocephala 146, 148–9, **149**
Cyperaceae (sedges) 167
Cytinus hypocistis 93, **93**

Dactylorhiza fuchsii 139
daffodils 121, **121**, **135**
daisy family see Asteraceae
Dalechampia 142
 D. dioscoreifolia **142**
damselflies 81, 208

Danaus plexippus 86
Daniellia pynaertii 72
dark-capped bulbul see *Pycnonotus tricolor*
Darwin, Charles 6, 36, **36**, 56, 69, 146, 156–7, 159, 171–2, 174, 180
 On the Origin of Species (1859) 36, 156, **157**, 166
Darwin, Erasmus 49
Darwin's orchid see *Angraecum sesquipedale*
Datura 121
 D. stramonium **121**
Daucus carota 122, **122**
Davidia involucrata 111
dawn redwood see *Metasequoia glyptostroboides*
daylily see *Hemerocallis*
dead-horse arum see *Helicodiceros muscivorus*
deception strategies **119**, 122–5, **124**, 130, **131**, 139, 142–4
deciduous plants 62
deforestation 201
deformed wing virus 206
Deherainia smaragdina 109
Dendrobium sinense 119, **119**
Derelomus chamaeropsis 118, **118**
desiccation 20
despeciation 157, 159
Diamond, Jared 203, 204
Diascia barberae **140**
diffraction graftings 108–9
Digitalis 27, 169
 D. canariensis **159**
 D. lutea **168**
 D. purpurea **159**
 D. × valinii **159**
dimethyl disulphide 115, 124
dioecious plants 36–7, **37**, 39, 46, 167
 angiosperms 61
 gymnosperms 55, 59
 and water pollination 51
diploid 16, **17**
Diptera (true flies) 94, 113, 123
disease resistance 11
dispersal potential 211
distyly 41
diurnal animals 170, 172, 173, 209
Dixon, Kingsley 211
Dobzhansky, Theodosius 156, **156**
dogbane see Apocynaceae
dogwood see *Cornus canadensis*
Dominican Republic 94
dormouse **73**
dotted loosestrife see *Lysimachia punctata*
Dracunculus vulgaris 187
dragonflies 81, **161**, 208
Drakea 122
 D. livida **123**
Drepanotermes 94, 105
duckweed see Lemnoideae
dune thistle see *Cirsium pitcheri*
Dunnett, Nigel 214
durian see *Durio zibethinus*
Durio zibethinus 192
dutchman's pipe plants see Aristolochia
dwarf palm see *Chamaerops humilis*

EBI fungicides 204
Echium wildpretii 100, **100**, 101, 111, **132**, 133

echolocation 74, 103, **103**, 124
ecological resilience 190, 202, 211
ecological specialisation 176-7, 181, 211
ecology 102, 185, 189-90
economics 184, 185
ecosystem facilitators 192
ecosystem services 185, **185**, 189, 192, 196
  cultural 185
  provisional 185
  regulating/maintenance 185
ecosystems 45, 46, 196-7, 200, 203, 211
eelgrass **47**, 50, **50**, 51
egg 12, 14, **15**, 16-17, 43
  and dioecious plants 36-7
  growth of the pollen grain towards 33-5, **33**
  and inbreeding avoidance strategies 41
  and wind pollination 57
Eichhorn, Susan 157
elaiophores 141
elaiosome 93
elbow orchids *see Arthrochilus*
Elodea
  E. canadensis 48, **48**
  E. nuttallii 50
elyptra (modified forewings) 89
embryos 10, 16, 23, 137
  abortion 36, 59
Encephalartos ferox 212
endotherms 73
energy stores 23
English yew *see Taxus baccata*
Ensifera ensifera **136**
Entomyzon cyanotis 80, **80**
Eonycteris spelaea **69**, 75
Ephedra 55, 113
  E. funerea **113**
epiparasitic species 105
epiphytes 24, 37-9, 134, 176
epithelial elaiophores 141
Equisetum (horsetail) 113
Ericaceae 27
erosion, soil 189
Erysimum scoparium 108, **108**
Erythrina crista-galli **172**
ethers, aromatic 120
Eucalyptus **67**, 142
euglossine bees 142, 175, 190, **190**
Eulophia parviflora **163**
Euphorbia 39, 53, 111, **111**, 133, 134, **134**
  E. abyssinica **166**
  E. aleppica 106, **106**
  E. awashensis **112**
  E. baylissii **112**
  E. dendroides 169, **169**
  E. guiengola **112**
  E. mafingensis **112**
  E. mellifera 158, **158**
  E. neococcinea **112**
  E. pulcherrima 111
  E. quadrilatera **112**
  E. stygiana 158, **158**, 212
  E. × pasteurii 158, **158**
Euphorbiaceae (spurge family) 93, 166
European Union (EU) 208
evil quartet 204
evolution 12, 36, 155-6, 159, 166, 176, 184, **192**

advergent 180, 181
convergent 50, 133, 166, 179, 180, 181
parallel 166
and pollination 180
predictability of 166
speed of 180
evolutionary ecology 102
evolutionary trees (phylogenies) **12**, 163, 165
exaptations 12, 22
Exesipollenites 85
exine 21, 51, 60
extinction 163, 201-2, 204, 208, 212, **212**, 213
  chains of 203
eyes 107

Fabaceae (bean family) 103, 171, 175
Faegri, Knut, *The Principles of Pollination Ecology* 165
farming *see* agriculture
female anatomy of the flower **11**, 21
female cones 57-9, **59**, 68, 84, 85, 116, **133**, 212
female gametophytes 33
Fenster, Charles 162, 176
fern allies 13
ferns 12, **13**, 14-17, **14**, 16, 43, 53, 133, **133**
fertilisers, nitrogenous 204
Ficus 147, 152, **152**
  F. hispida **152**
  'fruits' 94
fig wasp 94, 152, **152**
figs *see Ficus*
filaments 18, 26, 61
flavones **108**
flavonoids 108
flax *see Phormium*
Fleming, Theodore 150
flies 6, 71, 80, 85, 119, 174
  carrion 174
  fossilised 85, **184**
  fruit 94
  house 94
  hoverflies **6**, 81, 94, **97**
  true 94, 113, 123
floral tissues 131
flower colour 107-13, **107-9**
  and bird-pollinated plants 172, **172**
  chemicals of **108**
  colour changes 100-1, **100**, 124, 149, **149**
  and insect pollinated plants 174, 175
  and nocturnal species 171, 173
  and pollination syndromes 180
flower visitors
  non-pollinating 161, **161**
  see also animal pollination
flowerpeckers 76
flowers 18
  anatomy **11**, 21
  before foliage 104, **104**
  female anatomy **11**, 21
  patterns 101, 107
  petaloid 110
  pollen-only 139
  scent 114-20
  shape 101, 105, **105**, 175, 180
  signalling with 100-1
  size 101, 106-9
  specialisation 176, 177

symmetry 105
  three-dimensional structure 101
  unisexual 40
  and wind-pollinated angiosperms 61-2, **61**, **62**
forests **202**
  fragmentation **204**
fossils 10, **10**, 19, 77, 85, 133, 184
  amber 85, **85**, 94, **94**, **184**
foxglove *see Digitalis*
French lavender *see Lavandula stoechas*
French meadow-rue *see Thalictrum aquilegiifolium*
Freycinetia 139
  F. baueriana 208
fruit flies 94
fungal infections 142
fungi 11, 20, 105
fungicides 204
fungus gnats 94, 210

gametes 14-17, **15**, **17**, 23
  see also egg; sperm
gametophytes 14, 17-18, 21-2, 33
gaping dutchman's pipe *see Aristolochia ringens*
gardens 214, **214**
Garrya elliptica 61
gaseous exchange 12
geckos 71, 113
gene flow 163
gene pools 15
generalists 176-7, 209, 211
generative cells 18, **20**
genes 11, 16, 17, 36, 157
  and self-incompatibility 41
genetically modified (GM) crops 208
genetics 157, 159
Geranium (geranium) 27, 112
germination
  pollen grain 21, 33-5, **33-5**, 51, 137
  spore 14
Ghazoul, Jaboury 203
giant feather grass *see Stipa gigantea*
giant honey flower *see Melianthus major*
giant saguaro cactus *see Carnegiea gigantea*
giant water lily *see Victoria amazonica; Victoria cruziana*
giant white bird of paradise *see Strelitzia nicolai*
Ginkgo 57, **57**
  G. biloba **33**
Gladiolus **179**
  G. liliaceus **179**
Glaucis hirsutus **78**
Global Strategy for Plant Conservation 210
globeflowers *see Trollius*
glucose 139
Glycine max 194, **195**
glyphosate 201
GM (genetically modified) crops 208
Gnetales 85, 164, 167
Gnetum 55
Goldblatt, Peter 179, **179**
golden chalice vine *see Solandra maxima*
golden digger wasp *see Sphex ichneumoneus*

golden oats *see Stipa gigantea*
Gonepteryx rhamni **210**
gorse *see Ulex europaeus*
Gossypium (cotton) 193, 194, **195**, 208
Gould, Stephen 12, 166, 184
gourds (Cucurbitaceae) 19
Grab, Heather 209
Grant, Verne 162
Grant-Stebbins model 162-4
grapes 194, **195**
grasses *see* Poaceae
grasslands, frequently burned **64**
Great Lakes 212
great tit *see Parus major*
greenhouse gases 208
Grevillea **80**
Grotewold, Erich 122
guano 204
guelder-rose *see Viburnum opulus*
gum trees *see Angophora; Corymbia; Eucalyptus*
gums 142
Gurney, Dorothy Frances 187
Gymnadenia 88
  G. odoratissima **88**
gymnosperms 17-18, 21, 25, 160-1, 163-4, 167, 184
  and animal pollination 55-7, 59, 84-5, 89
  and inbreeding 36, 37
  and nectar production 133, **133**
  and pollen grain deposition 30
  and pollen grain germination 33, **33**
  and pollination drops 113, **113**
  sexual strategies of 59
  wind pollination in 55-9, **133**

Habenaria glaucifolia 88
Haber process 204
habitat, and wind pollination 63-4, **63-4**
habitat destruction 204, **204**
habitat fragmentation 204, **204**, 205
habitat rehabilitation 214
habitat restoration 210-11, 213
Habropoda laboriosa 193
Hamamelis mollis 187, **187**
hammer orchids *see Drakea*
handkerchief tree *see Davidia involucrata*
haploid 16, **17**, 18
Harcourt Arboretum 210
Hawai'i 206
hawkmoths 86-8, **87**, 91, 102, 136, 212
hay fever (allergic rhinitis) 21, 194
hay meadows 204
hazel *see Corylus avellana*
heart-shaped tongue orchid *see Serapias cordigera*
heat production 117, **117**, 134, **134**, 146, **146**, 148-9
heath *see Ericaceae*
hedgerows 211
hedges 189, **189**, 204, 210
Helianthemum 139
Helianthus 115
Helicodiceros muscivorus 124, **124**
Heliconius hecale 88, **88**
Helleborus (hellebores) 110, **110**, 134, **134**, 141, 146
  H. foetidus 137, 146
  H. niger **117**

INDEX 219

Helophilus trivittatus **97**
Hemaris thysbe **82**
Hemerocallis **137**
Henchie, Stewart 212
herbicides 208
heterotherms 86, 91
Himalayan balsam see *Impatiens balsamifera*; *Impatiens glandulifera*
HIPPO 204
Hitchmough, James 214
Holcoglossum amesianum 37, 38, **38**, 39, 41
Holland, Nathaniel 150
holly see *Ilex aquifolium*
holometabolous 83
Homo sapiens 10, 17
honey 90, **90**, 92, 193, **193**, 196
honey bee see *Apis mellifera*
honey bush see *Richea scoparia*
honey spurge see *Euphorbia mellifera*
honeyeaters 76, **77**, 80, **80**, 171
honeysuckle see *Lonicera periclymenum*
hornets 6, 119, **119**, **176**
hornworts 12, **52**, **53**
horsetails see *Equisetum*
house flies 94
hoverflies **6**, **81**, 94, **97**
human impact 183–97, 200, 213
human settlements, spread 204
hummingbird clearwing see *Hemaris thysbe*
hummingbird hawkmoth see *Macroglossum stellatarum*
hummingbirds see Trochilidae
Hyacinthoides non-scripta 190, **190**
Hyacinthus (hyacinth) **199**
hybrid species 158
Hydrilla verticillata 54, **54**
Hydrostachys 47
Hymenoptera **83**, 164
Hypericum 139

Ilex aquifolium 188
Impatiens
  I. balsamifera 207
  I. glandulifera 92
inbreeding 14, 36–42, 205
  advantages 41–2
  avoidance strategies 36–41, 62
inbreeding depression 12, 36, 38, 59
indoles 115
industrial applications 22
inflorescences **10**
Insect Pollinators Initiative 214
insect-dispersed pollen grains 24
insecticides 208
insectivores 76, 78
insects 81, 131
  eyes 107
  nocturnal 86, **87**, 88, 173
  preservation in amber 85, **85**, 94, **94**, **184**
  and rewards 136, 138
  and speciation 160, 162, 163–4
  see also specific types of insect
Intergovernmental Science-Policy Platform on Biodiversity and Ecosystem Services (IPBES) 201
  Pollination, Pollinators and Food Production (2017) 186, **186**

International Pollinator Initiative 186
International Union for Conservation of Nature, Red List assessments 202
intine 23
invasions 204, 206–8
invertebrate pollinators 81–3, 85–95
  conservation 201
  and pollination syndromes 172–6
  see also specific types of invertebrate pollinators
IPBES see Intergovernmental Science-Policy Platform on Biodiversity and Ecosystem Services
Iris (iris) 175

Jacaranda mimosifolia 104, **104**
jade vine see *Strongylodon macrobotrys*
jasmine tobacco see *Nicotiana alata*
Jimsonweed see *Datura stramonium*
Johnson, Steven 115, 162, 176, 178
jojoba see *Simmondsia chinensis*
Joshua tree see *Yucca brevifolia*
Juncaceae (rushes) 167

kākā see *Nestor meridionalis*
kākābeak, New Zealand see *Clianthus puniceus*
Kalmia (kalmias) 29
kangaroo paw see *Anigozanthos flavidus*
kapok see *Ceiba pentandra*
ketones 115, 141
kiekie see *Freycinetia baueriana*
king protea see *Protea cynaroides*
Knox, Bruce 49
Kolkata 214
Kölreuter, Joseph 156, **156**
Krameria 141
  K. bicolor **141**
Kremen, Claire 202
Kunz, Thomas 192

La Notte, Alessandra 185, 189
labellum 122, 123, 142, 144, 180
Lacerta dugesii **70**
Lack, Andrew, *The Natural History of Pollination* 165, **165**
Lagarosiphon 49
Lapeirousia 163
  L. jacquinii **163**
large blue butterfly see *Phengaris arion*
large garden bumblebee see *Bombus ruderatus*
large tiger hoverfly see *Helophilus trivittatus*
Larix (larch) 58, **58**
latex 169
Lathyrus odoratus 105, **105**
Lavandula stoechas 118, 122
'leaky phloem' hypothesis 132–3, 135
learning 101
leaves 135
  colourful 111
Lemnoideae (duckweed) 46, **46**, 48
lemurs 73, **73**
Lepidoptera (butterflies and moths) 86–8, 113, 123

see also butterflies; moths; specific species
Leptonycteris
  L. curasoae **75**
  L. yerbabuenae **138**
Leptospermum scoparium 193, **193**
lesser knapweed see *Centaurea nigra*
lesser long-nosed bat see *Leptonycteris yerbabuenae*
lifecycle length 160
light pollution 209, **209**
Lilium (lily) 26, 27, 111
  L. candidum 114, **114**
linalool 120, 123
Linnaeus, Carl 112
Lithophragma 147
liverworts 12, **15**
lizards 69, 70–1, **70**, 169, **169**
  see also geckos
Lobelia 82
London Zoo, Butterfly Paradise 188
long-tailed sylph see *Aglaiocercus kingii*
Lonicera
  L. fragrantissima 114
  L. periclymenum **129**, 172, 173
lords-and-ladies see *Arum maculatum*
lorikeets 76
lycopods 12
Lysichiton camtschatcensis **107**
Lysimachia punctata 141, **141**

mānuka see *Leptospermum scoparium*
mānuka honey 193, **193**
Macaca sinica **72**
Macroglossum stellatarum 98, **116**, 172, 173
Macrozamia lucida **85**, 116
Maculinea arion see *Phengaris arion*
Madagascar 73
Madeira wall lizard see *Lacerta dugesii*
Madonna lily see *Lilium candidum*
Magnolia (magnolia) 110
  M. grandiflora 104
Mahonia (mahonia) 29
maiden silvergrass see *Miscanthus sinensis*
maize see *Zea mays*
malachite sunbird see *Nectarinia famosa*
male cones 58, 59, **59**, **85**, 95, 116
mallow family see Malvaceae
Malus domestica (apple) 194, **195**
Malvaceae (mallow family) 192
malvin 108
mammals 72–5, **74–5**, 180, 190, 210
  conservation 202
  nocturnal 80
  non-flying 170
  and pollination syndromes 180
  see also specific types of mammals
Manduca sexta 121
Manning, John 179, **179**
mantises 81
Marcgravia evenia 103, **103**
Marcgraviaceae 103
marginal aquatic plants 48
marigold **22**
marsh marigold see *Caltha palustris*
Masarinae (pollen wasps) 85, 94, 137

mason bees see *Osmia*
Matthews, Philip 148
Mauritius 113
May, Robert, Baron May of Oxford 81, 215
Mayr, Ernst 157, **157**
McCormick, Molly 214
meadow-rues see *Thalictrum*
meadows
  hay 204
  traditional flower 210
Mead's milkweed see *Asclepias meadii*
Mediterranean-type regions 165, **165**
Megaselia fly 105
meiosis 17, **17**
Melaleuca uncinata 105
Melianthus major 113, **115**
Meligethes aeneus **23**
Meliphagidae (honeyeaters) 76, **77**, 80, **80**, 171
memory 88, 90, 91, 117
Metasequoia glyptostroboides 58, **58**
methylglyoxal 193
midges 139, **139**
  see also biting midges
milkweeds see Asclepiadaceae
mimicry 180
Miscanthus sinensis 187, **187**
mistletoe see *Viscum album*
mites 15, 206, **207**
mitosis **17**
mock orange see *Philadelphus coronarius*
Mocuna holtonii 103
monarch butterfly see *Danaus plexippus*
monkey puzzle tree see *Araucaria*
monkeys 72, **72**
monoecious plants 167
  angiosperms 62
  and inbreeding avoidance strategies 37, 39, 40
  and wind pollination 54, 59
monolectic relationships 119
monoterpenes 120
Monticola (rock thrushes) 178
moon, phases of the 113
Morgan's sphinx moth see *Xanthopan morganii*
mosses 12, 15, **15**, 43
moth-pollinated flowers 27
moths 17, 85, 86–8, **87**, **129**, 179, 209, **209**
  attraction strategies 116
  diurnal 101
  and pollination syndromes 172–3, **173**
  and rewards 136, **150**
Mucuna holtonii 103
Musa 75
mutation 36, 160
mutualism 90, 127, 147, 150, 159, 172, 178, 180, 185
ß-myrcene 116
Myrmica sabuleti 213

nailwort see *Paronychia argentea*
Narcissus 121, **121**
naringenin 22
National Aeronautics and Space Administration (NASA) 155

natural capital 196–7
natural history 156
natural sciences 214
natural selection 55, 69, 155, 157, 166
nature
    and mankind 184, 185
    putting a price on 196–7
    threats to 204–9
Nature (journal) 109, 196, 201
navigation 30
nectar 67, 68, 73, 74, 75, 210
    absence of 130, 139
    amino acid content 133, 136
    appeal 132–3
    attracting animal pollinators with 101–2, 108, **108**, 113, 174–6, 179
    and bat pollination 171
    and bird pollination 76, 78, 172
    bitter 178
    caffeine content 117
    colour 113, 178
    composition 136–7
    defensive function 133
    energy content 132, 136, 137
    fermentation/alcoholic 134
    fructose-rich 136, 137
    glucose-rich 136, 137
    and insect pollinators 83, 86–7, **86**, 90–4, **90**, **92**, **94**, 100, 103, 150, 153, 173
    and moth pollinators 173
    nicotine-laced 102, **102**
    in non-flowering plants 133
    as pollinator reward 131, 132–7
    production 135
    scent of 101, 119, 120
    sucrose-rich 101, 136, **136**, 137
    sugar content 101, 132–3, 135–7, **136**, 179
    volume of 136–7
nectar guides 121, 175
nectar robbers 76, 132, 135, 172, 178
nectar spurs 132
nectaries 112, **112**, 131, 137
    and heat production 117, **117**, 134, **134**, 146
    hidden 132, **132**
    and lizard pollination 169
    location 134, **134**
    modified 141
    in non-flowering plants 133
Nectarinia famosa 77
Nectariniidae (sunbirds) 76, **77**, 79, **79**, 178
nectarless flowers 130, 139
neonicotinoids 204, 208
Neozeleboria cryptoides 180
Nestor meridionalis **171**
'new synthesis' 157
New Zealand 208
New Zealand flower thrip see *Thrips obscuratus*
New Zealand kākābeak see *Clianthus puniceus*
Nicotiana (tobacco plants) 102
    N. alata **102**
nitrogen 137
nitrogen-fixing leguminous plants 189
Niveoscincus microlepidotus 71, **71**
nocturnal species 113, 170, 172, 209, **209**
    insects 80, 86, **87**, 88, 173

mammals 80
moths 80
Norway spruce see *Picea abies*
nursery provision 147–51
Nuttall's pondweed see *Elodea nuttallii*
Nymphaea lotus 48, **48**
Nymphaeaceae (water lilies) 110, **110**

oak see *Quercus*
Ochroma pyramidale 192
(E)-ß-ocimene 116, 118, 123
oils 140–1, **140**, **141**
olfactory signals 99, 114–25, 141, 146, 150, 152
olfactory skills 88, 92
olive sunbird see *Cyanomitra olivacea*
Ollerton, Jeff 108, 162, 176, 177, 184, 201–2
Ophrys apifera 180
oranges 194, **195**
Orchidaceae (orchid family) 6, **24**, 29, 32, 88, **88**, 93–4, 105, 190, **190**
    and attracting animal pollinators 101, 122–3
    and deception strategies 141–4, **144**, **145**
    epiphytic 24, 37–9, 134
    rewards produced by 139, 146
    scent 119, **119**
    and speciation 163, **163**, 164, **164**
    see also specific species
orioles 76
Osmia **188**
osmosis 12
ostiole 152
ovary 11, 26, 29, 134, 151–2, 160
overexploitation 204
ovule 26, 29
    and avoidance of inbreeding 36–7, 39, 41
    and deposition of pollen grains 32
    and insect pollinators 100
    and parasites 147
    race to the 33
    water pollination 53
    of yuccas 151
Oxford 158
oxidative stress 87, 117

Pachycereus 147
    P. schottii 150–1, **150**
painted lady see *Vanessa cardui*
painted petal irises see *Lapeirousia*
Palaeomyopa **184**
palms 46, 147
    see also specific species
Papaver 139
Papilio machaon **69**
parasitism 93, **93**, 152, 185, 206
    of pollen grains 147
    of seeds 147, **147**, 150–2, **150**
parlour palm see *Chamaedorea elegans*
Paronychia argentea 93, **93**
Parrotia persica 111, **111**
parrots 69
Parus major 208
Pasteur spurge see *Euphorbia × pasteurii*

Patagonian giant bumblebee see *Bombus dahlbomii*
pathogens 206
pedunculate oak see *Quercus robur*
Penstemon 87
peony 22
perception space 107
perianth 54, 62, 110, **110**, 167
peroxides 87
Persian ironwood see *Parrotia persica*
pesticides 204, **209**
petaloid flowers 110
petals 11, 39, 134
petunia **100**
Phaethornithinae 78
Phelsuma gecko 113
Phengaris arion 213, **213**
phenols 178
pheromones 88, 99, 180
    mimicry 115, 119, 122–3
Philadelphus coronarius 187, **187**
Philodendron
    P. martianum 146
    P. solimoesense 146
philosophy 185
phloem 135
Phormium **77**
phosphorus 137
photosynthesis 135, 160, 185
Phylloscopus collybita 77
phylogenies see evolutionary trees
Picea abies **56**
pigments 108, **108**–9
'pin' state 40–1, **40**
pink-spotted sphinx moth see *Agrius cingulata*
Pinus (pine) 26, 59, **59**
    P. sylvestris **57**
Pirone-Davies, Cary 59
pistils 11, 46, **46**
plant kingdom 11
plant life history 17
plant species problem 157, 159
Plantaginaceae 168
Plantago (plantain) 55, **55**, 169
    P. lanceolata **168**
plastids 108
Platanthera leucophaea 212, **212**
plum yew see *Cephalotaxus*
plume thistle see *Cirsium rivulare*
plunger mechanism 39, 39
Poaceae (grasses) 24, **24**, 45, **60**–1, 61–2, **160**, 164, **164**, 167, **167**, 187, 210
poinsettia see *Euphorbia pulcherrima*
pollen baskets 91, **91**
pollen beetle **23**, **69**, 83
pollen brushes 89
pollen coat 20–3, **21**, 30, 51, 137
pollen dusting 196–7
pollen grains 43, 160, 161
    air sacs 57, **57**
    of *Aloe vryheidensis* 178
    amino acid content 138
    and animal pollination 24, 67–8, 74, **74**, 79–80, 83, 85, 90–1, **90**–**2**, 93–4, 100–1, 103, 139, 172, 174, **174**
    appearance 18, 19
    and attracting pollinators 111, 124
    and beetle pollination 174, **174**
    and bird pollination 172
    boat-shaped 57

buoyancy tanks **20**
buzz collection 91, 139
colour 100, **100**, 101, 111
composition 137
conspecific 34, 35, 39
costs of **131**
definition 18
deposition 30–2
and dioecious plants 36–7
dormant 27
energy content 132
evolution 10
of figs 152, **152**
filamentous 50
fossilised 10, **10**
gathering 139
germination 21, 33–5, **33**–**5**, 51, 137
of the giant water lily 148, 149
and inbreeding avoidance strategies 36–7, 39–41
and insect pollinators 24, 85, 90–1, **90**–**2**, 93–4, 100
interior **21**, 23
and invertebrate pollination 83
longevity 24–5
and nectar scent 120
and oil 141
overproduction 137
parasites 147
partially hydrated 33
presentation 28–9
protection 111
protein content 137, 138
release/transportation 9, 17, 26–8, **26**, **27**, 30, 58, 62, 120, **133**
as reward 131, 137–9
rod-shaped 50, 51
and self-fertilisation 42
and self-pollination 42
of the senita cactus 150
spherical 50
and stigmatic exudates/secretions 139
structure 20–1, **20**–**1**
surface-to-volume ratio 137
threats to 20
and water pollination 48–51, **51**, 53, 65, 167
and wind pollination 46–7, 54–62, **55**, **56**, **60**, 64–5, **133**, 167
of yuccas 151
see also pollinia
pollen screens **26**
pollen spoons 89
pollen tubes 21, 33–4, **33**–**4**, 41, 53, 60, 137–8
pollen wasps see Masarinae
pollen-only flowers 139
pollenkitt 21, 60
pollination 183–97
    abiotic 46–7
    aesthetic value of 187, 188
    biological significance of 155–81
    biotic 46–7
    costs 128–9, 146
    critical importance of 184, 186
    cultural importance of 188
    definition 9–43
    ecological value of 189–90
    economic cost of 194
    economic value of 193–4, 196–7
    and evolution 180
    fossil records of 184
    future of 214

hand 196-7, **196**
as mutualism 127
and release of the pollen 26-8
and speciation 158, 160-4
and species diversity 160-5
stages of 26-35
value of 184-5, 187-90, 193-4, 196-7
see also animal pollination; rewards of pollination; self-pollination; water pollination; wind pollination
pollination ecology 102
pollination (pollen) drops 33, 56, 58, 59, 113, **113**, 133
pollination syndromes 159, 165-81, 211
abiotic 167, 169
and bats 192
contemporary existence 176-7
debate summary 178-9
and invertebrate pollinators 172-6
types of 167, 169-75
and vertebrate pollinators 169-72
pollinator failure 42
pollinator rewards *see* rewards of pollination
pollinator shifts 176-7, **177**, 179, 180
pollinia **24**, 119, 123, 134, 144, **144**, **145**
pollution 204, 208-9
polyene tail **108**
Polyommatus icarus 9
poppies *see* Papaver
population growth 204
Port Jackson willow *see* Acacia saligna
Portugal 118
possums 69, 170, **170**
potato family *see* Solanaceae
Potentilla 139
prairies, temperate **63**
Prance, Ghillean 148
pre-adaptation 12
Primula (primula) 40, **40**
proboscis 85, 86, **86**, 116, 121, **121**, 172
Proctor, Michael, *The Natural History of Pollination* 165, **165**
Prodoxidae (yucca moths) 30, 147, 150, 151, **151**
proline 133, 136, 138
Prosthemadera novaeseelandiae **77**, **171**
protandry 39
Protea 73, 120
P. cynaroides **180**
P. welwitschii 120, **120**
Proteaceae (protea family) 180
prothallus 14, 15, 16, **16**, 17
protogyny 39, 148
Prunus dulcis (sweet almond) **182**, 194, **195**
Pteridium aquilinum **16**, 133
Pterogodium catholicum 205, **205**
purpletop vervain *see* Verbena bonariensis
Puya 171
Pycnonotus tricolor **178**
Pycnontidae (bulbuls) 178
pyrethroids 204

Quercus 62
Q. robur 61, **61**
temperate 64
tropical 64
Quisqualis indica 124, **124**

Rafflesia arnoldii 106, 107
Raguso, Robert 114, 115, 122
rainbow lorikeet *see* Trichoglossus moluccanus
Rangoon creeper *see* Quisqualis indica
Ranunculaceae 169
Ranunculus 105, **105**, 134
rapeseed **23**
Raven, Peter 157
Ravenala madagascariensis 73
reactive oxygen species 87
red angel's trumpet *see* Brugmansia sanguinea
Rediviva **140**
R. peringueyi 205
redundant features 122, **122**, 202
reed warbler (Eurasian) *see* Acrocephalus scirpaceus
reproduction
asexual 11, **11**, 46
see also sexual reproduction
reproductive isolation 157
resins 142
rewards of pollination 73, 95, 99-101, 105, 117, 119, 125, 127-53, 175
benefits of 128-9
costs of 128-9
food and water benefits 131-9, 153
and generalist flowers 177
non-food benefits 140-5, 153
and scent 119
rhinoceros beetles **137**
Rhizanthella gardneri 94, 105, **105**, 210
Rhododendron **27**
Richea scoparia 71, **71**
Ricinus communis **29**
rock thrushes *see* Monticola
rockroses *see* Cistaceae; *Cistus*; *Helianthemum*
Rosa 'Zéphrine Drouhin' 114
Rosas-Guerrero, Victor 176
rose chafer *see* Cetonia aurata
Rosmarinus officinalis (Rosemary) 118, **118**
rufous-breasted hermit *see* Glaucis hirsutus
rushes *see* Juncaceae
Rutidosis leptorrhynchoides 205, **205**

S-locus allele system 41, 205
sago cycad *see* Cycas revoluta
saguaro cacti *see* Carnegiea gigantea
St John's wort *see* Hypericum
Salvia (sage) 29
sand-tailed digger *see* Cerceris arenaria
saponarin 108
Sarcococca (sweet box) 114
Sardinian warbler *see* Sylvia melanocephala
Saxicolinae (chats) 178
scarab beetle *see* Cyclocephala
scarlet banksia *see* Banksia coccinea
scent 123, 141
attracting pollinators with 102, 107, 109, **109**, 148, **148**, 172-5, 190, **190**
beauty of 187
bee 98
and bird pollination 172
carrion-/faeces-mimicking 107, 109, 115, 124, 136, 174
changes in 100, 101
classification 115
of figs 152
flower 100, 101, 114-20
and moth-pollinated plants 172-3, **173**
and nectar guides 121
as reward 142-5
roles of 116-17
selection 119
source of 118-19
and warmth 117, 124, 146
Schiestl, Florian 115, 162
Schrödinger, Erwin 102, 185, 197
Scots pine *see* Pinus sylvestris
Scrophulariaceae 168
seagrass 48, 49, **49**, 50, 51
search vehicles (pollen aggregations) 50
seasonality 64
sedge *see* Carex
seed dispersal 93
seed ferns 10
seeds 50, 161
larval parasites 147, **147**, 150-2, **150**
overproduction 137
as reward 137
and wind pollination 46
self-fertilisation 12, 24, 32, 35, 36-42
and abortion of embryos 59
advantages 41-2
deliberate 37-9, **38**
ease of 42
reliability of 42
speed of 42
self-incompatible species 34, 41
self-pollination 21, 29
and abortion of embryos 59
and animal-pollinated plants 38, 64
deliberate 37-9, **38**
and a shortage of animal vectors 64
and wind pollination 54
self-recognition 41
senita cacti *see* Pachycereus schottii
senita moth *see* Upiga virescens
sensory ecology 102
sepals **11**, 39, 110, **110**, 134
'separation in space' strategies 40-1
'separation in time' strategies 39
Serapias
S. cordigera **128**
S. lingua 146
sexual reproduction 11-17, **11**, 43, 46
advantages of 36
costs of 128-9
in gymnosperms 59
problems with 11-12
Seymour, Roger 148
short-spurred fragrant orchid *see* Gymnadenia odoratissima
signallers 99
signals 99-125, **102**, 108

acoustic 99, 103, **103**
combined 101
complexity 100-1
message conveyed 102
olfactory 99, 114-25, 141, 146, 150, 152
physical 99, 121
visual 99, 104-13, 121-5, 150
Silene 10
Simmondsia chinensis 62, **63**
skatoles, iconic 115
small skipper butterfly *see* Thymelicus flavus
snakeflies 81
snapdragons *see* Antirrhinum
snow skink *see* Niveoscincus microlepidotus
soil, erosion 189
Solanaceae (potato family) 108
Solandra
S. grandiflora **106**
S. maxima 106
solitary bees 141
Sonoran Desert 150-1
Sorghum bicolor **160**
sori *see* sporangia
South Africa 163, **163**, 178, 180, 208
South African iris *see* Lapeirousia jacquinii
southeastern blueberry bee *see* Habropoda laboriosa
southern long-nosed bat *see* Leptonycteris curasoae
southern magnolia *see* Magnolia grandiflora
soybean 194, **195**, 208
spadix 107, **107**
Spanish bayonet *see* Yucca whipplei
spathe 107, 124
specialisation 176-7, 180-1, 211
speciation 11, 158, 160-5, 181
Grant-Stebbins model of 162-4
rates of 15
species diversity 11, **189**, 190, 202-3, **202**, 214
and bees 175, **175**
and pollination 160-5
reduction 204
see also biodiversity
sperm 12, 14-16, **15**, 17, 18, 22, 43
and deposition of pollen grains **32**
and inbreeding avoidance strategies 41
release from the pollen grain 33
and water pollination 46
Sphex ichneumoneus 162
Sphingidae 86
spiders 69
sporangia 14, **14**, 16-18, 29
spores 14, 16-18, **17**, 22, 25
sporophytes 16, 17
sporopollenin 14, 17, 21, 22, **22**, 43, 51
Sprengel, Christian 156, **156**
springtails 15, **15**
spurge family *see* Euphorbiaceae
spurs 132, **132**, **140**, 163
stamen **11**, 18, 29
and attracting pollinators 106, 111
and heat production 149
maturation 27
and nectaries 134
and protandry 39
and scent 121

and water pollination 49, 53
and wind pollination 46, **46**, **62**
of yuccas 151
staminodes 139, **139**
Stanhopea 144, 190, **190**
S. wardii **127**
Stapelia gigantea **130**
starch 23, 135, **135**, 148
Stebbins, G. Ledyard 162
Steenhuisen, Sandy-Lynn 120
Sternbergia clusiana **135**
stigma **11**, 26, 29, 148, 160-1
and animal pollination 68, 79, 91, 93, 111, 123, 169
and deposition of pollen grains 30, 32, **32**, 34, 39
and inbreeding prevention strategies 40-1, **40**
introduction of pollen grains to 34, **34**, 35, **35**, 39
lack of in gymnosperms 56
and lizard pollination 169
and scent 121
and self-pollination 42
of the senita cactus 150
and water pollination 49, 50, 51, 53, 65, 167
and wind pollination 54, 60, 62, **62**, 65
of yuccas 151
stigmatic exudates/secretions 131, 133, 139
stinking hellebore see *Helleborus foetidus*
Stipa gigantea **61**, 187
Stone, Graham 28
Strelitzia nicolai 79, **79**
striations 108-9
Strongylodon macrobotrys 108, **109**, 171, **171**
stylar processes 148, **149**
style 26, 41, 160
and attracting pollinators 111
of figs 152
and nectaries 134
and the presentation of pollen 29, **29**
and water pollination 53
and wind pollination 62
of yuccas 151
Stylidium 29
sugars 23, 101, 132-3, 135-7, **136**, 179
sulphur compounds 115
sunbirds see Nectariniidae
sunflowers see *Helianthus*
sustainability 214
swallowtail butterfly see *Papilio machaon*
swamp milkweed see *Asclepias incarnata*
sweet almond see *Prunus dulcis*
sweet box see *Sarcococca*
sweet pea see *Lathyrus Odoratus*
sweet violet see *Viola odorata*
sword-billed hummingbird see Ensifera
syconium 152, **152**
Sycopsis sinensis 111
Sylvia
 S. atricapilla 77
 S. melanocephala 77
Symphyotrichum 6

tachinid fly **28**

Tasmania 71
taxonomic impediment 69
Taxus 26
 T. baccata 37, **37**, 39
Tegeticula
 T. maculata 151
 T. synthetica 151
 T. yuccasella 151, **151**
Teilhard de Chardin, Pierre 156, **156**
temperature rises 209
 see also heat production
Tenerife 133
termites 94, **94**, 105
terpenoids 115
tetrapyrroles 108
Thalassia testudinum 49, **49**, 51, **51**
Thalassidendron ciliatum 50
thale cress see *Arabidopsis thaliana*
Thalictrum 139, 169
 T. aquilegiifolium **169**
Theobroma cacao 69, 94, 139, **139**
thermogenesis 117, **117**, 134, **134**, 146, **146**, 148-9, 180
Thomas, Jeremy 213
Thomisidae (crab spiders) 117, **117**
Thomson, James 162
thrips 46, 69, 85, 116, 137, 147
Thrips obscuratus 85
'thrum' state 40-1, **40**
Thymelicus flavus 88
thymine wasp see *Neozeleboria cryptoides*
Thymus (thyme) 213
 T. serpyllum **213**
thynnid wasp 122-3
tiger longwing see *Heliconius hecale*
titan arum see *Amorphophallus titanum*
tobacco hornworm moth see *Manduca sexta*
tobacco plants see *Nicotiana*
tongue orchid see *Serapias lingua*
toque macaque see *Macaca sinica*
tower of jewels see *Echium wildpretii*
trap-lining 72, 74, 78, 88, 90, 142
traveller's tree see *Ravenala madagascariensis*
tree spurge see *Euphorbia dendroides*
Trichoglossus moluccanus **69**
trichomes **140**, 141
Trichosurus vulpecula 208
Trochilidae (hummingbirds) 74, 76-9, 85, 101, 136, **136**
 hermit 78
 long-tongued 136
 non-hermit 78-9
Trochilinae 78
Troglodytes troglodytes 142
Trollius 147, **147**
true flies 94, 113, 123
tūī see *Prosthemadera novaeseelandiae*
Tulipa (tulips) 110, **129**
Tyria jacobaeae 172
tyrosine 138

ubiquitin pathway 34
Ulex europaeus **32**
ultraviolet colours 92, **92**, **107**, 109, 111, 121
ultraviolet light 12, 20, **107**

umbrella liverwort see *Marchantia polymorpha*
Umwelt concept 107
unilateral incompatibility 34
unisexual flowers 40
United Kingdom 214
United Nations (UN) 6, 186, **186**
Upiga virescens 150-1, **150**

vacuole 108
Vallisneria 49, 50, 54
 V. spiralis **47**
van der Niet, Timotheüs 162, 176
van der Pijl, Leendert, *The Principles of Pollination Ecology* 165
Vanessa cardui **155**
Vanilla planifolia **196**
Varecia variegata 73
Varroa destructor mite 206, **207**
vectors 28, 30, 43, 49
 see also animal pollination; water pollination; wind pollination
vegetative cells **20**
Verbena bonariensis 116, **155**, 173
vertebrate pollinators 70-80, 169-72
 see also specific vertebrate pollinators
Vespa crabro **176**
Vespula pensylvanica 206
vexillum 103
Viburnum (viburnum) 34
 V. opulus 89, **89**
 V. rhytidophyllum **35**
Victoria
 V. amazonica 139, 146, 148-9, **148-9**
 V. cruziana 148
Viola (violet) 134
 V. odorata 41, **41**
Viscum album 199, **199**
vision 109
 colour vision 87, 92
visual signals 99, 104-13, 121-5, 150
Vitis vinifera cultivars 194, **195**
Vogel, Stefan 140, 162
volatile organic compounds (VOCs) 115, 119, 150
volatiles 19
von Uexküll, Jakob 107
Vrba, Elizabeth 12

waggle dance **68**
Wardhaugh, Carl 210
Ward's stanhopea see *Stanhopea wardii*
warty hammer orchid see *Drakea livida*
Waser, Nikolas 162, 176
wasp orchid see *Chiloglottis*
wasps 71, 85, 90, 94, **94**, 122-3, 135, 176, **176**
 and conservation issues 204
 extinction 201-2
 parasitic 152
 thynnid 122-3
 see also specific types of wasps
water 12, 33
 loss 12
 supplies 135
water lilies see *Nymphaea lotus*; Nymphaeaceae
water pollination 46, 48-53, 65

at the water's surface 49-50
by aquatic animals 49
and pollination syndromes 167
submarine 51-3
weevils 118, **118**
Welwitschia 55
Western columbine see *Aquilegia formosa*
western pygmy possum see *Cercartetus concinnus*
western underground orchid see *Rhizanthella gardneri*
western yellowjacket wasp see *Vespula pensylvanica*
white asphodel see *Asphodelus albus*
white fringed orchid see *Platanthera leucophaea*
white rhatany see *Krameria bicolor*
white-eyes see Zosteropidae
Wielandiella angustifolia 10
wild carrot see *Daucus carota*
wildfire 64
Williamsoniaceae **10**
Willis, Kathy 202
Wilson, Edward 204
wind pollination 45-7, 54-65, 163, 176-7, 184, 187, 190
 and angiosperms 46, 60-2, 64
 and animal pollination 28, 46, 47, 54-5, 169, **169**, 176-7
 and crops 193
 and gymnosperms 55-9
 and habitat 63-4, **63-4**
 and marginal aquatic plants 48
 and pollination syndromes 167-9
 popularity 46
 and seasonality 64
 and speciation 164, **164**
 and wind-dispersed pollen 14, 24, 30, 43, 62
wings, butterfly/moth 86
winter-flowering honeysuckle see *Lonicera fragrantissima*
Wolffia 46, **46**
woodland stars see *Lithophragma*
woodlands
 temperate 63, **63**, 64, **64**
 tropical 63-4
woody plants 160
Wytham Woods, Oxford 208

Xanthopan morganii 172
xylem 135

yeast 134
Yeo, Peter, *The Natural History of Pollination* 165, **165**
yew see *Taxus*
Yucca (yucca) 30, 147, 150, 151
 Y. brevifolia 151
 Y. whipplei 151
yucca moths see Prodoxidae

Zea mays 62, **62**, 167
zingiberene **115**
zoology 214
Zostera **50**
 Z. marina 50, 51
Zosteropidae (white-eyes) 178
Zygaenidae (burnet moths) 172
zygomorphic blooms 105, **105**
zygotes 16, 17, **17**